Lecture Notes in Computer Science 831
Edited by G. Goos and J. Hartmanis

Lecture Notes in Computer Science 821

Edited by G. Goos and J. Hartmanis

Advisory Board: W. Brauer D. Gries J. Stoer

Vincent Bouchitté   Michel Morvan (Eds.)

Orders, Algorithms, and Applications

International Workshop ORDAL '94
Lyon, France, July 4-8, 1994
Proceedings

Springer-Verlag
Berlin Heidelberg New York
London Paris Tokyo
Hong Kong Barcelona
Budapest

Series Editors

Gerhard Goos
Universität Karlsruhe
Postfach 69 80
Vincenz-Priessnitz Straße 1
D-76131 Karlsruhe, Germany

Juris Hartmanis
Cornell University
Department of Computer Science
4130 Upson Hall
Ithaca, NY 14853, USA

Volume Editors

Vincent Bouchitté
Michel Morvan
Laboratoire de l'Informatique du Parallélisme, Ecole Normale Supérieure de Lyon
46 Allée d'Italie, F-69364 Lyon Cedex 07, France

CR Subject Classification (1991): F.2.2, G.2.1, G.2.2, G.1.6

ISBN 3-540-58274-6 Springer-Verlag Berlin Heidelberg New York
ISBN 0-387-58274-6 Springer-Verlag New York Berlin Heidelberg

CIP data applied for

Printed in Germany

Typesetting: Camera-ready by author
SPIN: 10472649 45/3140-543210 - Printed on acid-free paper

Preface

This volume contains the texts of the survey papers and of the research articles which have been presented at "ORDAL'94: Orders, Algorithms and Applications".

Ordered sets and more specifically algorithmic aspects of order theory have taken a growing importance in computer science. In 1987, Ivan Rival organized the first conference on "Algorithms and Order" in Ottawa, Canada. Since then, a number of new developments have been made and algoritmics of ordered structures enjoys a recognized place in computer science as in mathematics. We hope that ORDAL'94 will serve its purpose as an opportunity for scientific exchanges and cooperation among specialists of all concerned domains (algorithmics, order theory, applications to computer science). ORDAL'94 could then become the first of a series of regular meetings on the subject.

Among the 35 papers submitted, 9 have been accepted for presentation and publication, after a selection based on high scientific criteria. The choice was difficult, and we want to thank all the contributors for their interest in ORDAL'94.

Editing this issue would not have been possible without the work of the program committee members and the help of the many referees whose detailed reports allowed us to select the appropriate papers. We are grateful to them.

In the present volume, research contributions come after survey papers which correspond to the talks given by invited speakers. We thank them for their participation. The success of the manifestation is largely due to their recognized competence on the subject. We also want to acknowledge the following invited speakers who have presented talks which are not reflected by written papers: U. Faigle, H. Kierstead, R. H. Möhring, M. Pouzet, R. Tamassia, R. Wille and P. Winkler.

We also thank the many who helped in all aspects of this meeting.

Lyon, France, May 1994 V. Bouchitté, M. Morvan

Préface

Ce volume contient les textes des exposés de synthèse et des articles de recherche qui ont été présentés à "ORDAL'94: Ordres, Algorithmes et Applications".

Les ensembles ordonnés et plus spécifiquement les aspects algorithmiques de la théorie de l'ordre ont pris une place de plus en plus importante en informatique. En 1987, Ivan Rival a organisé à Ottawa au Canada la première conférence sur le sujet intitulée "Algorithms and Order". Depuis, un grand nombre d'avancées ont été faites et l'algorithmique des structures ordonnées occupe une place incontestable en informatique comme en mathématiques. Nous espérons que ORDAL'94 jouera son rôle qui est avant tout de donner aux spécialistes des différents domaines concernés (algorithmique, théorie de l'ordre, applications à l'informatique) l'occasion de se rencontrer et de travailler en collaboration. ORDAL'94 pourrait alors devenir la première édition d'une série de rencontres sur le sujet.

Parmi les 35 articles soumis, 9 ont été acceptés pour présentation et publication, après une sélection basée sur des critères élevés de qualité scientifique. Le choix a été difficile et nous tenons à remercier tous les auteurs pour l'intérêt qu'ils ont porté à ORDAL'94.

L'édition de ce document n'aurait pas été possible sans le travail des membres du comité de programme et l'aide des nombreux arbitres dont les rapports détaillés nous ont permis de sélectionner les meilleurs articles. Nous les en remercions ici.

Dans le présent volume, les travaux de recherche sont précédés d'articles de synthèse qui correspondent aux exposés des conférenciers invités. Nous les remercions pour leur participation. Le succès de cette manifestation est sans aucun doute dû en grande partie à leur compétence reconnue sur le sujet.

Nous remercions aussi tous ceux qui ont apporté leur aide ou leur soutien à l'organisation de ce colloque.

Lyon, France, mai 1994 V. Bouchitté, M. Morvan

Program committee

V. Bouchitté
M. Habib
R. Jégou
R. H. Möhring
M. Morvan
M. Pouzet
J.-X. Rampon
I. Rival

Supported by

Laboratoire de l'Informatique
du Parallélisme

Ecole Normale Supérieure
de Lyon

MINISTERE DE
L'ENSEIGNEMENT SUPERIEUR
ET DE LA RECHERCHE

Human Capital and Mobility: *DIMANET*

Table of Contents

Invited Contributions

Contributed Papers

Table of Contents

Invited Contributions

Contributed Papers

Bit-Vector Encoding for Partially Ordered Sets

Michel Habib and Lhouari Nourine

Département d'Informatique Fondamentale
LIRMM, Université Montpellier II — CNRS UMR C09928
161 rue Ada, 34392 Montpellier Cedex 5, FRANCE.
(email: (habib,nourine)@lirmm.fr).

Abstract. Given a lattice L, we propose a tree representation of L. We show that this tree contains a bit-vector encoding of L and then how to compute from this tree the lattice operations (meet and join). Algorithms which provide bit-vectors encodings for partial orders have been recently proposed in the literature. Given a partial order P we recall that computing an optimal bit-vector encoding of P is NP-Complete. From a theoretical lattice point of view we propose bit-vector encodings and study their optimality. We end by suggesting a data structure (lazy MacNeille completion) which can have many applications.

Keywords: Lattice theory, Galois lattice, MacNeille Completion, Distributive Lattice and Lattice of Ideals, Bit Vector Encoding, Bounded Dimension.

1 Introduction and Motivations

In this paper we try to bring together results obtained in lattice theory: Bouchet [6], Habib *et al.* [17], Markowsky [21, 22, 23], Morvan and Nourine [25], Wille [30], and questions emerging from applications in computer science. In fact from very practical issues in Databasis: Agrawal *et al.* [1], Ellis [10], in Artificial Intelligence when representing knowledge via hierarchies or conceptual graphs: Godin [15], Ellis and Lehmann [11], in programming languages development: Aït-Kaci *et al.* [2] and Caseau [7], in distributed systems: Mattern [24] and Charron-Bost [8], the question of efficient representations for acyclic digraphs, orders or lattices has been raised.

Until now, most of the theoretical work for representing orders is related with various notions of dimension see Trotter [29], or Gambosi *et al.* [12, 13, 14]. But unfortunately these works produce mainly NP-complete parameters. So it remains a very challenging question for this new discipline of algorithmic order theory : **What classes of orders can be efficiently represented** ? By efficient we mean a representation using minimal memory, few preprocessing time and that allows quick answer to reachability queries ($x \leq y$?).

In this paper we first study this representation problem for lattices, since via MacNeille completion or ideal-completion the lattices play a central role for orders. Based on theoretical results a tree representation is produced for any lattice and its properties are studied. We exhibit the relationships between bit-vector encodings as defined and used by Aït-Kaci *et al.* [2], Caseau [7], Ellis and Lehmann [11], with order theoretic parameters, such as bounded dimension.

Our approach produces natural bit-vector encodings and a lot of research directions for new heuristics. So we end by a proposal for a lazy MacNeille completion of an order, i.e. a data structure which is contained in the classic MacNeille completion and which can be update when needed. For example if the meet of two elements is asked, and if it does not already exist in the structure, we simply add it at its right place.

2 Definitions and Notations

A partially ordered set $P = (X, \leq_P)$ is a reflexive, asymmetric and transitive binary relation on a set X. We denote by $<_P$ the strict ordering associated with P. When necessary, we may consider P as a directed graph (X, E) where $E \subset X^2$ and $(x, y) \in E$ iff $x \leq_P y$. Two distinct elements x and y are said to be comparable if $x <_P y$ or $y <_P x$. Otherwise they are said incomparable (denoted by $x||y$). We say that y covers x (denoted by $\prec$) iff $x <_P y$ and there is no z such that $x <_P z <_P y$; if y covers x then x is an immediate predecessor of y.

A subset A of X is called an *antichain* (resp. *chain*) of P if it contains only pairwise incomparable (resp. comparable) elements. The height (resp. width) of P, denoted by $height(P)$ (resp. $width(P)$) is the size of a maximal chain (resp. antichain) in P.

Let $P = (X, <_P)$ be a poset, we define the following sets of an element x of P by $Pred(x) = \{y \in X \mid y <_P x\}$ the set of predecessors of x, $Succ(x) = \{y \in X \mid x <_P y\}$ the set of successors of x, $\downarrow x = \{y \in X \mid y \leq_P x\}$ the ideal corresponding to x and $\uparrow x = \{y \in X \mid x \leq_P y\}$ the filter corresponding to x.

An element $z \in X$ is an upper bound of $x, y \in X$ if $x \leq_P z$ and $x \leq_P z$. The element z is called the least upper bound or *join* of x and y if $z \leq_P t$ of all upper bounds t of x and y. The greatest lower bound or *meet* is defined dually. We denote $x \vee y$ (resp. $x \wedge y$) the least upper bound (resp. greatest lower bound) of x and y. A non-empty ordered set P is called a lattice if $x \vee y$ and $x \wedge y$ exist for all $x, y \in X$. It is clear that a finite lattice has one minimal element and one maximal element denoted respectively by *bottom* and *top*. Since we are dealing with algorithms the lattices we considered are supposed to be finite. For a recent book on lattice theory the reader is referred to Davey and Priestley [9].

Let L be a lattice. An element $x \in L$ is said to be join-irreducible if $x \neq$ *bottom* and $x = y \vee z$ implies $x = y$ or $x = z$ for all $y, z \in L$. In other words x covers only one element. Meet-irreducible elements are defined dually. We denote the set of join-irreducible elements of L by $J(L)$ and the set of meet-irreducible elements by $M(L)$.

We denote by $I(P)$ the set of all antichains of P ordered as follows: $A <_{I(P)} B$ if and only if for all $x \in A$ there is $y \in B$ such that $x \leq_P y$. It is well known that $I(P)$ such ordered is a distributive lattice (see. Birkhoff [4]). We denote by $AM(P)$ the suborder of $I(P)$ (which is a lattice generally non distributive) restricted to the antichains of P maximal relatively to inclusion.

Let L be a finite distributive lattice. It is well known that the induced order of join-irreducible element is isomorphic to the meet-irreducible one. Moreover the order $J(L)$ or $M(L)$ is a representation of L. By representation we mean that L can be reconstructed as the ideal lattice of $J(L)$ or $M(L)$ [4].

For non-distributive lattice both join-irreducible and meet-irreducible elements are needed. That is any lattice L can be characterized by the bipartite order $Bip(L)$ computed as follows from L : $Bip(L) = (M(L), J(L), <_{Bip(L)})$ where $x <_{Bip(L)} y$ if and only if $x \in M(L)$, $y \in J(L)$, and $x \not\geq_L y$.

The lattice L can be reconstructed as the lattice of maximal antichains of $Bip(L)$ denoted by $AM(Bip(L))$ or the Galois lattice of the complementary relation of $Bip(L)$.

The following theorem can be easily deduced from [3, 21, 22, 23, 25, 27, 30].

Theorem 1. *Let* $L = (X, E)$ *be a lattice. Then* $L \cong AM(M(L), J(L), \not\geq_L) \cong AM(X, J(L), \not\geq_L) \cong AM(M(L), X, \not\geq_L)$.

The above result can also be directly shown by using the operation of reduction of a bipartite order which keeps invariant the lattice of maximal antichains. This operation is detailled in section 5. So there exist three natural ways to represent a lattice by means of a maximal antichains lattice. For sake of simplicity let us stick to the second one, and define $BIP(L) = (X, J(L), \not\geq_L)$.

3 Tree Encoding of Lattices

In this section we give a tree representation of any lattice and we consider how classical operations on lattices can be computed on this tree.

Given a lattice $L = (X, <_L)$ and $BIP(L) = (X, J(L), \not\geq_L)$ its associated bipartite order. The successors of each vertex x in X of $BIP(L)$ form the code of x denoted by $code(x)$.

Lemma 2. *The function* $x \longrightarrow code(x)$ *is one-to-one and* $x <_L y$ *iff* $code(x) \supset code(y)$.

Proof. From Morvan and Nourine [25], we know that the lattice of maximal antichains (denoted by $AM(BIP(L))$) of $BIP(L)$ is isomorphic to L, therefore there exists a lattice isomorphism $\varphi : L \longrightarrow AM(BIP(L))$. Now suppose that $x \neq y$ in L such that $code(x) = code(y)$, necessarily $\varphi(x) = \varphi(y)$ which yields a contradiction.

For the second part of the lemma let us suppose that $x <_L y$. Then $\downarrow x \subset \downarrow y$ and $L \backslash \downarrow x \supset L \backslash \downarrow y$. Therefore $(L \backslash \downarrow x)/J(L) \supset (L \backslash \downarrow y)/J(L)$. The result follows since $code(x) = (L \backslash \downarrow x)/J(L)$. Suppose now that $code(x) \supset code(y)$. Clearly $x \notin code(x)$ then $x \notin code(y)$. Thus $x <_L y$ since $y \not\geq_L x$.□

Lemma 3. *Let L be a lattice and $x, y \in L$, then $code(x \vee y) \subset code(x) \cap code(y)$ and $code(x \wedge y) \supset code(x) \cup code(y)$.*

Proof. Since we have $x <_L x \vee y$ and $y <_L x \vee y$, then $code(x \vee y) \subset code(x)$ and $code(x \vee y) \subset code(x)$. The result can be completed dually. □

Let τ be a linear extension of $J(L)$ and we sort the vertices of $J(L)$ according to τ. Let $CODE(x)$ denote the decreasing list of elements in $code(x)$ according to τ. Using these codes a tree can be recursively constructed as follows:

Start with a root denoted by *top* with $code(top) = \emptyset$, and for a given vertex x in L, its unique father in the tree is the vertex z such that $CODE(z)$ is lexicographically maximal among all vertices satisfying $code(z) \subset code(x)$.

An edge xy of the tree is labelled with $CODE(x) \setminus CODE(y)$, and therefore from the root to any vertex of the tree, the labels are decreasing according to τ. Such a tree denoted by $T(L)$ is called *a tree representation* for L, and its height is bounded by $|J(L)|$, since the size of a maximal code is bounded by $J(L)$ and edges are labelled at least by singletons. Let us denote by Δ the maximum degree of $T(L)$, which is at most $|L|$. See Figure 1 for an example of such a tree.

Theorem 4. *let $L = (X, E)$ be a lattice given by its transitive reduction graph, with $|X| = n$ and $|E| = m$ there exists a tree representation $T(L)$ computed in $O(n*m)$, which allows reachability computations in $O(|J(L)|)$ and $O(\Delta*|J(L)|)$ for computations of meet and join operations.*

Proof. To compute the tree from L, given by its transitive reduction, we first compute $BIP(L)$ in $O(n * m)$, and then compute a linear extension τ of $J(L)$. This can be done using the degrees in $BIP(L)$, without computing explicietely the suborder $J(L)$. To complete the tree, we begin with the root and insert successively the vertices as leaves, taking the vertices ordered by their codes.

Now we suppose to have in our data structure a direct access from each vertex of the lattice to its place in the tree. Using Lemma 1 reachability can be simply obtained by testing inclusion of two ordered lists of size at most $|J(L)|$ and thus can be done in $O(|J(L)|)$. Similarly when computing $x \vee y$, $A = code(x) \cap code(y)$ can be obtained in $O(|J(L)|)$. To find $x \vee y$, we just have to start from the root of the tree down to the first vertex z such that $code(z) \subseteq A$. This search can be done in $O(\Delta * |J(L)|)$. □

Generally this tree $T(L)$ is not unique (since it depends on the selected linear extension τ). A lattice L is called *simplicial* if its height is equal to $|J(L)|$ or $|M(L)$, also called join-extremal or meet-extremal by Markowsky [23]. $T(L)$ is a spanning tree of L if L is a distributive or a simplicial lattice. But it is wrong in general and we know examples for which no tree representation spans L.

Furthermore if L is distributive all its tree representations have their edges labelled by singletons. For simplicial lattices we only know the existence of at least one such tree. Moreover we know how to compute such a tree [19].

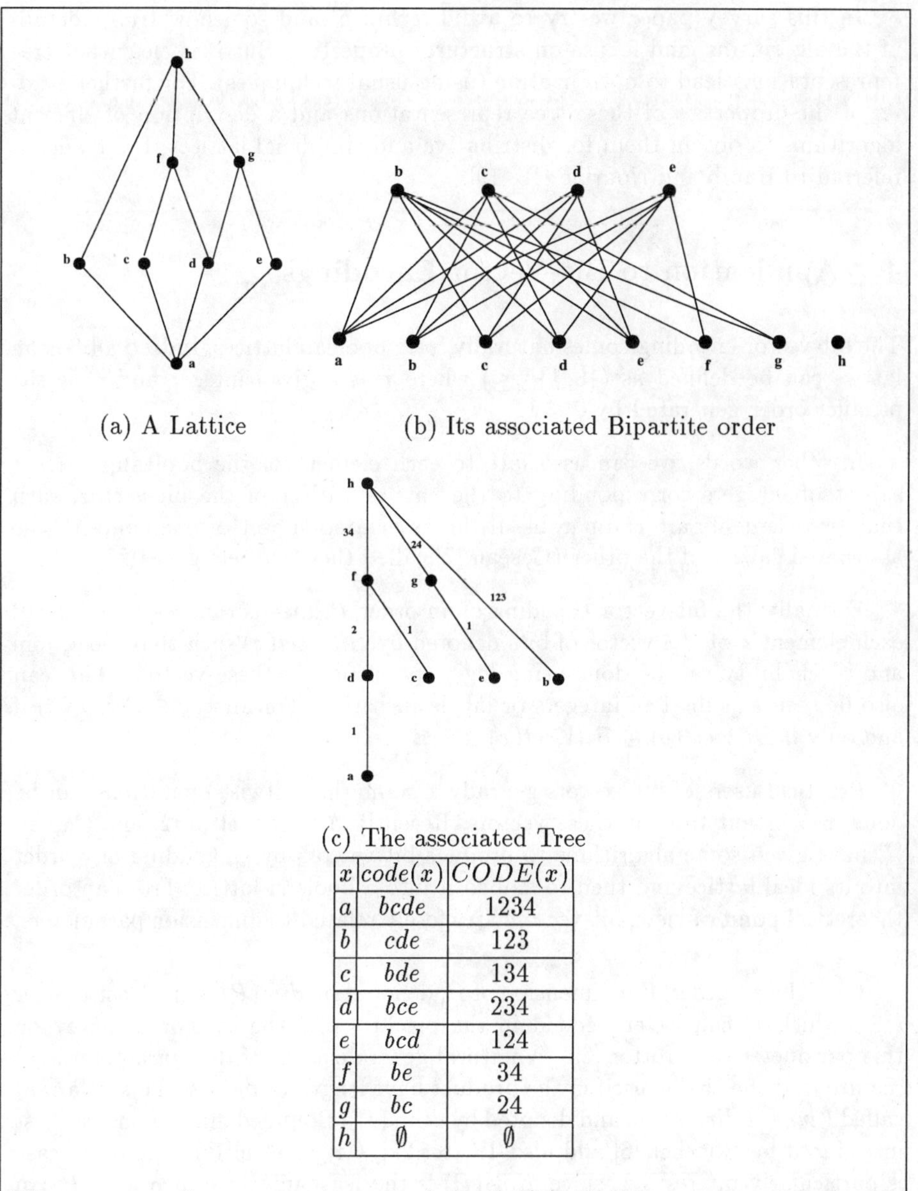

x	$code(x)$	$CODE(x)$
a	$bcde$	1234
b	cde	123
c	bde	134
d	bce	234
e	bcd	124
f	be	34
g	bc	24
h	$\emptyset$	$\emptyset$

We compute a linear extension $\tau = dceb$ of $J(L)$. This linear extension can be computed according to the number of predecessors in $BIP(L)$.

Fig. 1. Example of a tree representation and its associated encoding for a lattice

In this survey paper we try to avoid technical and somehow tricky details of the algorithms, and focuse on structural properties. But it is clear that tree representations lead to optimization (using usual techniques). For further studies of the properties of these tree representations and a description of efficient algorithms to obtain them for distributive and simplicial lattices, the reader is referred to Habib and Nourine [18, 19].

4 Application to Bit-Vector Encodings

The bit-vector encoding comes naturally from boolean lattices. Indeed a boolean lattice can be defined as $(\{0,1\}^n, \leq)$ where n is a given integer and $\leq$ is the product order generated by $0 \leq 1$.

In other words, we can associate to each element of the boolean lattice a subset of integers corresponding to the entries with 1 of the bit-vector, such that two elements are comparable if the associated subset of one contains the associated subset of the other (Caseau [7] called these subsets genes).

Formally, the bit-vector encoding of an order P, means that we associate to each element x of P a vector of bits denoted by $BitVect(x)$ such that meet, join and reachability can be done using logic operations on these vectors. This can also be seen as subset of integers or labels such that: For all $x, y \in P, x <_P y$ if and only if $BitVect(y) \subset BitVect(x)$.

Practical users of bit-vectors genrally assume that bitwise operations can be done in constant time on these vectors. Recently Ait-Kaci et al [2] and Caseau [7] have given some algorithms to produce bit-vectors by embedding any order into its ideal lattice and then to embed it into a boolean lattice. From an order theoretical point of view, bit-vectors are closely related to dimension parameters.

Let P be an order, the dimension of P, denoted by $dim(P)$ is the least integer t for which P can be embedded in the product of t chains. For a survey on this parameter see Trotter [29]. A natural generalization of this parameter is to require that the chains used in the product have length at most k. This invariant called *bounded dimension* and denoted by $dim_k(P)$. Bounded dimension, was first introduced by Bouchet [6] and also Trotter[29], Griggs et al[16]. The dim_2 case is particularly interesting, since $dim_2(P)$ is the least integer t such that P can be represented by a family of subsets of Q, ordered by inclusion, or the smallest size of the boolean lattice on which you can embed P. Bit-vector codings and dim_2 are closely related, since $dim_2(P)$ can be interpreted as the least integer t such that P admits a sized t bit-vector encoding. So $dim_2(P)$ is the optimal size of a bit-vector encoding for P. Unfortunately computing $dim_2(P)$, therefore optimal bit-vectors is already NP-hard as shown by Stahl and Wille [28].

Theorem 5. *For every integer k, $dim_k(P) = dim_k(DM(P))$.*

Proof. This result is implicitely contained in Bouchet's work [6], since he proved that if P can be embedded into a complete lattice L, so does its MacNeille completion. A product of t chains of length k is a complete lattice, which gives the result. □

We notice that this Bouchet's result does not require P to be finite.

4.1 Bit-vector Encoding of Lattices

Let $L = (X, E)$ be a lattice. Then L is isomorphic to maximal antichains of the bipartite order $BIP(L) = (M(L), J(L), \not\geq)$.

Proposition 6. *Let $L = (X, E)$ be a lattice. Then L can be encoded with a bit-vector of size $min(|M(L)|, |J(L)|)$.*

Proof. As we already used for the construction of the tree representation $T(L)$, one can easily obtain a code associated to each vertex namely $code(x)$, it suffices to transform it into a bit-vector of size $|J(L)|$. Since we could have start with $(M(L), X, \not\geq)$, then the result follows. □

Lemma 7. *Let $L = (X, E)$ be a lattice. If L is a chain then L can be encoded optimaly with the size of the bit-vector encoding equal to $|M(L)| = |J(L)| = |L| - 1$.*

Proof. Clearly the subset of the maximal element of L is empty since the associate subset of all other elements contains the empty set. Let $L = x_1, x_2, ..., x_{n-1}, x_n$ the linear order. Thus $BitVect(x_n) =$ and since $x_{n-1} <_P x_n$ then $BitVect(x_{n-1})$ is different from $BitVect(x_n)$ and must contain it. Then let $BitVect(x_{n-1}) = \{n-1\}$. Suppose now that $BitVect(x_i) = \{n-1, n-2, ..., i\}$ is optimal and let us compute the optimal code of x_{i-1}. It is clear that the code of must be different from the code of $x_i, ..., x_{n-1}$ and x_n and contains all their code. Thus we must add an integer to have the code of x_{i-1}. □

Theorem 8. *If $L = (X, E)$ is distributive or simplicial then L can be encoded optimaly with a bit-vector of the size equal to its heigth.*

Proof. By proposition 6, we conclude that L can be encoded with a bit-vector of the minimal size between $M(L)$ and $J(L)$. If L is simplicial or distributive then its height is $|J(L)|$ or $|M(L)|$. Since there exist a chain in L of length $|J(L)| + 1$, so using lemma 7 the size of the bit-vector is optimal. □

Remark 1: The theorem 8, shows us that embedding a partial order into a simplicial lattice instead of a distributive lattice is sufficient to ensure an optimal bit-vector encoding.

For a non-simplicial lattice the encoding problem can be tranformed into an embedding problem into a simplicial lattice having minimal height.

4.2 Embedding an order into its MacNeille Completion

The MacNeille completion L of P is the smallest lattice in which you can embed L. Such an embedding is provided by the order embedding $\varphi : P \longrightarrow L$ with $\varphi(x) = \downarrow x$,

Theorem 9. *$dim_2(P) = dim_2(DM(P))$.*

Proof. Suppose that $DM(P)$ has a Bit-vector encoding at most size k. Since P is a suborder of $DM(P)$ then the size of an encoding of P is k. Now suppose that P has a bit-vector encoding of size k. Then P can be represented by $(P, 1...k, \not\geq_P)$ such that $x \not\geq i$ iff i belongs to the code of x. So the maximal antichain of $(P, 1...k, \not\geq)$ is a lattice L with P as a suborder of L. Using the proposition 6 the size of the encoding of L is less or equal to k. Since $DM(P)$ is a suborder of L the result follows. □

We notice that this yields, in the finite case, another proof of the case $k = 2$ of theorem 5.

Therefore an optimality condition can be deduced for bit-vector encodings of an order P. Indeed if $DM(P)$ is simplicial or distributive our algorithm in the following section produces an optimal bit-Vector. It seems interesting to further study class of orders for which optimal bit-vectors are known.

5 Lazy MacNeille Completion of a partial order

Let $P = (X, <)$ be a partial order and $DM(P)$ its Dedekind-MacNeille completion. Clearly each element $x \in P$ has an image $\varphi(x) \in DM(P)$. Moreover each join-irreducible or meet-irreducible element of $DM(P)$ has a reverse image in P. Since any lattice can be characterized using only join-irreducible and meet-irreducible elements, thus we can compute the tree representation of $DM(P)$ restricted to the elements of P. This means that all elements added to P to obtain $DM(P)$ are not irreducible elements. This result is based on the following theorem:

Theorem 10. *[25] $DM(P) \cong AM(X, X, \not\geq_P)$.*

Now let us show how to determine the join-irreducible elements of $DM(P)$.

First, we associate to $P = (X, <_P)$ the bipartite order $BIP'(P) = (X, X_1, E_1)$ such that X_1 is a copy of X and $(x, y) \in E_1$ iff $x \in X$, $y \in X_1$ and $x \not\geq_P y$.

5.1 Reduction of a bipartite order

Definition 11. Let $BIP'(P) = (X, X_1, E_1)$ the associated bipartite order, and $x \in X_1$. Then x is deleted from X_1 if the following property holds.

$$Pred(x) = \bigcup_{i=1}^{k} Pred(x_i) \text{ with } x_i \in X_1 \qquad (1)$$

The obtained order is the reduced bipartite order as defined in [25], denoted by $RED(BIP'(P)) = (X, X_2, E_2)$.

The induced order of join-irreducible elements of $DM(P)$ are given by the following proposition:

Proposition 12. *[25] Let $P = (X, E)$ be a partial order and $RED(BIP'(P)) = (X, X_2, E_2)$ its reduction. Then the set $\{Pred(x), x \in X_2$ in $RED(BIP'(P))$ ordered by inclusion is somorphic to $J(DM(P))$.*

The dual result can be stated using the reduction of X instead of X_1. Using the proposition 12, its dual and the theorem 1 we obtain the following result.

Theorem 13. $DM(P) \cong AM(X, X_2, \not\geq)$.

Now the set of join-irreducible elements of $DM(P)$ is X_2, and X is a subset of elements in $DM(P)$. The following section will show how we can associate a tree to $DM(P)$ containing only elements of P.

5.2 A subtree of $DM(P)$

Clearly P is a suborder of $DM(P)$. That is if $DM(P) = (V, U)$ then $X \subseteq V$. As for lattices the successors of the elements in X in $RED(BIP'(P))$ correpond to their code. We compute a linear extension of $\{Pred(x), \subset\}, x \in X_2$, and sort the code of elements in X according to the linear extension. As a consequence we can use the result of section 3 to compute a subtree of $DM(P)$. But in general we have to create some dummy vertices to keep the tree structure.

Let $x, y \in P$ such that

(a) $code(x) \cap code(y) = code \neq \emptyset$
(b) For all $i \in code(x)$, $j \in code(y)$, $k \in code$ we have $i < k$ and $j < k$.

If there does not exist $z \in P$ such that $code(z) = code$, then we create a dummy vertex z with this code. For example here z may represent the least upper bound of x and y in $DM(P)$.

Theorem 14. *The number of dummy vertices created is at most $|X|$.*

Proof. An easy proof by induction shows that when inserting a new vertex with its code in the already built tree, it has a unique inserting point (because of the selected linear extension). This point is added as a leaf, but in some cases we have to attach it between two vertices. Then we creat a dummy vertex.

In the whole at most $|X|$ dummies are created.□

Corollary 15. *The obtained tree is a subtree of $DM(P)$ which contains at most $2 * |X|$ nodes.*

This subtree representation of $DM(P)$ may have a lot of applications. For example if P is a crown then $DM(P)$ has a size $2^{|P|}$; so we cannot compute the whole $DM(P)$. Thus the use of this subtree and complete this subtree if necessary while the execution of the program. That is if we want to compute the least upper bound of x and y, we test if there exist, otherwise, we add a new vertex to the tree. This method corresponds to a lazy completion of P. Moreover the bit-vector encoding is contained in this tree.

Furthermore, in some applications, the programmers need only an order in between P and $DM(P)$, i.e. they really add vertices which have some semantics in the application.

6 Conclusion

In this paper we have produced two related representations for lattices. The first one, based on trees, provides easy computations for reachability, meet and join. The second one is a bit-vector encodings that provides reachability in O(1). We have shown how to derive bit-vectors for any orders.

By analogy with distributive lattices for which the tree representation yields algorithms having the best known complexity, see Habib and Nourine [18, 19, 26], we are wondering if one can derive good algorithms for generating the lattice of maximal antichains or the MacNeille completion lattice of a given order as needed in some applications, see Godin and Mili [15] or Caseau [7]. These algorithms have to be compared with previous works for Galois lattices Bordat [5], for maximal antichain lattices Jard *et al.* [20].

References

1. Rakesh Agrawal, Alex Borgida, and H.V. Jagadish. Efficient management of transitive relationships in large data bases, including is-a hierarchies. *ACM SIGMOD*, 1989.
2. Hassan Aït-Kaci, Robert Boyer, Patrick Lincoln, and Roger Nasr. Efficient implementation of lattice operations. *ACM Transactions on Programming Langages and Systems*, 11(1):115–146, january 1989.
3. G. Behrendt. Maximal antichains in partially ordered sets. *Ars Combin.*, C(25):149–157, 1988.
4. G. Birkhoff. *Lattice Theory*, volume 25 of *Coll. Publ. XXV*. American Mathematical Society, Providence, 3rd edition, 1967.
5. J.P. Bordat. Calcul pratique du treillis de gallois d'une correspondance. In *Math. Sci. Hum, 96*, pages 31–47, 1986.
6. A. Bouchet. Codages et dimensions de relations binaires. *Annals of Discrete Mathematics 23, Ordres: Description and Roles, (M. Pouzet, D. Richard eds)*, 1984.
7. Yves Caseau. Efficient handling of multiple inheritance hierarchies. In *OOPSLA'93*, pages 271–287, 1993.
8. B. Charron-Bost. *Mesures de la Concurrence et du Parallélisme des Calculs Répartis*. PhD thesis, Université Paris VII, Paris, France, Septembre 1989.

9. B. A. Davey and H. A. Priestley. *Introduction to lattices and orders.* Cambridge University Press, second edition, 1991.
10. G. Ellis. Efficient retrieval from hierarchies of objects using lattice operations. In *Conceptual Graphs for knowledge representation, (Proc. International conference on Conceptual Structures, Quebec City, Canada, August 4-7, 1993), G. W. Mineau, B. Moulin and J. Sowa, Eds, Lecture Notes in Artificial Intelligence 699, Springer, Berlin,* 1993.
11. G. Ellis and F. Lehmann. Exploiting the induced order on type-labeled graphs for fast knowledge retrieval. In *Proc. of the 2nd International conference on Conceptual Structures, August 16-20, 1994), College Park, Maryland, Lecture Notes in Artificial Intelligence, Springer-Verlag, Berlin,* 1994.
12. G. Gambosi, J. Nesetril, and M. Talamo. Efficient representation of taxonomies. In *TAP-SOFT, CAAP Conf. Pisa,* pages 232–240, 1987.
13. G. Gambosi, J. Nesetril, and M. Talamo. Posets, boolean representations and quick path searching. In *Proc. of the 14th International colloque on Automata, Languages and Programming, Lecture Notes in Computer Science 267, Springer-Verlag, Berlin,* 1987.
14. G. Gambosi, J. Nesetril, and M. Talamo. On locally presented posets. *Theoretical Comp. Sci.,* 3(70):251–260, 1990.
15. R. Godin and H. Mili. Building and maintening analysis-level class hierarchies using galois lattices. In *OOPSLA'93,* pages 394–410, 1993.
16. J.R. Griggs, J. Stahl, and W.T. Trotter. A sperner theorem on unrelated chains of subsets. *J. Comb. theory. (A),* pages 124–127, 1984.
17. M. Habib, M. Morvan, M. Pouzet, and J.-X. Rampon. Extensions intervallaires minimales. *C. R. Acad. Sci. Paris,* I(313):893–898, 1991.
18. M. Habib and L. Nourine. A linear time algorithm to recognize distributive lattices. Research Report 92-012, submitted to *Order,* LIRMM, Montpellier, France, March 1993.
19. M. Habib and L. Nourine. Tree structure for distributive lattices and its applications. Research report, LIRMM, Montpellier, France, Avril 1994.
20. C. Jard, G.-V. Jourdan, and J.-X. Rampon. Computing on-line the lattice of maximal antichains of posets. Thechnical report, IRISA, Rennes, France, February 1994.
21. G. Markowsky. Some combinatorial aspects of lattice theory. In Houston Lattice Theory Conf., editor, *Proc. Univ. of Houston,* pages 36–68, 1973.
22. G. Markowsky. The factorization and representation of lattices. *Trans. of Amer. Math. Soc.,* 203:185–200, 1975.
23. G. Markowsky. Primes, irreducibles and extremal lattices. *Order,* 9:265–290, 1992.
24. F. Mattern. Virtual time and global states of distributed systems. In M. Cosnard and al., editors, *Parallel and Distributed Algorithms,* pages 215–226. Elsevier / North-Holland, 1989.
25. M. Morvan and L. Nourine. Sur la distributivité du treillis des antichaînes maximales d'un ensemble ordonné. *C.R. Acad. Sci.,* t. 317-Série I:129–133, 1993.
26. L. Nourine. *Quelques propriétés algorithmiques des treillis.* PhD thesis, Université Montpellier II, Montpellier, France, June 1993.
27. K. Reuter. The jump number and the lattice of maximal antichains. *Discrete Mathematics,* 1991.
28. J. Stahl and R. Wille. Preconcepts of contexts. *in Proc. Universal Algebra (Sienna),* 1984.

29. W.T. Trotter. *Combinatorics and Partially Ordered Sets: Dimension Theory.* Press, Baltimore. John Hopkins University, 1992.
30. R. Wille. Restructuring lattice theory. *in Ordered sets,I. Rival, Eds. NATO ASI No 83, Reidel, Dordecht, Holland*, 1982.

Intervals and Orders: What Comes After Interval Orders?

Kenneth P. Bogart*

Dartmouth College, Hanover, NH, 03755, U.S.A.

Abstract. In this paper we survey two kinds of generalizations of the ideas of interval graphs and interval orders. For the first generalization we use intervals in ordered sets more general than the real numbers. For the second generalization, we restrict ourselves to intervals chosen in the real numbers, but we define two vertices to be adjacent (in the graphs) or incomparable (in the orders) only when the intervals overlap by more than a specified amount. Each of these generalizations suggests new avenues for research.

1 Introduction

A graph is called an *interval graph* if for each vertex x there is an interval I_x in the real numbers with the property that $I_x \cap I_y \neq \emptyset$ if and only if there is an edge between x and y. Interval graphs arise naturally as models in subjects as apparently unrelated as anthropology and electrical engineering. (See the book by Golumbic [16] for a discussion of some applications. See the book by Fishburn [15] or Trotter [31] for a much more complete survey of the theory than is given here.)

The first theorem we might find in the subject is that the complement of an interval graph is a comparability graph. In other words, if G has vertex set X and edge set E, if we use $\binom{X}{2}$ to denote the 2 element subsets of X, and if G' is the graph with vertex set X and edge set $E' = \binom{X}{2} - E$, then G' has a transitive orientation. This means that we can construct a transitive relation R (of the same size as E') which has either (a, b) or (b, a) (but not both) whenever $\{a, b\} \in E'$. Since the relation of a graph is irreflexive as well, R will be the strict order relation of a (partial) ordering. How do we define this ordering? Define $x < y$ (i.e. $(x, y) \in R$) to mean that I_x lies entirely to the left of I_y. This relation is clearly irreflexive and transitive, and so it is a strict (partial) order. The orders that arise in this way are called *interval orders*. For later reference it is useful to note that we might also have defined $x < y$ to mean that all points of I_x are less than or equal to all points of I_y. (The difference is that now two comparable intervals may have an endpoint in common.) The family of orders we get by applying this definition to sets of intervals of positive length is again the family of interval orders, even though a given set of intervals could could give two different orderings from the two different definitions.

* Supported by ONR Contract N0014-91-J1019

We have seen that an interval graph is the complement of a comparability graph. However, not all cocomparability graphs (complements of comparability graphs) are interval graphs. For example, the cycle on four vertices is the complement of a union of two disjoint 2-chains (See Figure 1). Thus it is a cocomparability graph

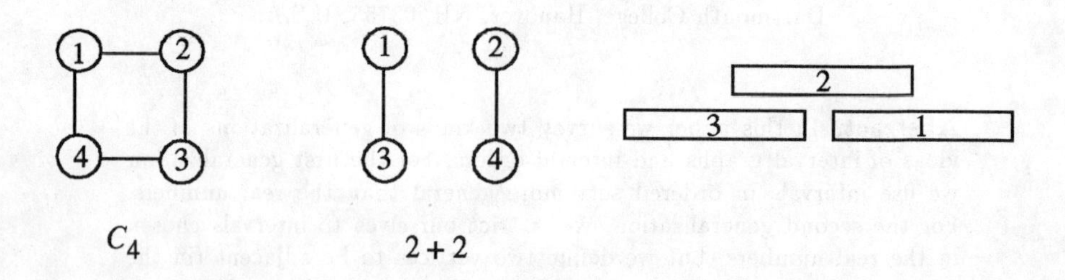

Fig. 1. The complement of C_4 is orientable as $2 + 2$, and neither is representable.

As we see in Figure 1, if we orient the complement so that vertex 3 is below vertex 1, then the interval representing 3 (shown as a rectangle) must be to the left of the interval representing 1, and the interval representing 2 must overlap both of these, leaving nowhere to put an interval that overlaps the intervals representing vertices 1 and 3 but not the one representing vertex 2.

Perhaps surprisingly, if an ordered set has no restriction to four elements isomorphic to the ordered set denoted as $2+2$ in Figure 1, it is an interval order. To show this to be the case, we would need to determine endpoints for intervals and decide which vertices have which endpoints. Let us proceed by example. We use a diagram such as Figure 2 to describe an ordering; to get the ordering from the diagram, direct all edges upwards and take the transitive closure of the digraph that results.

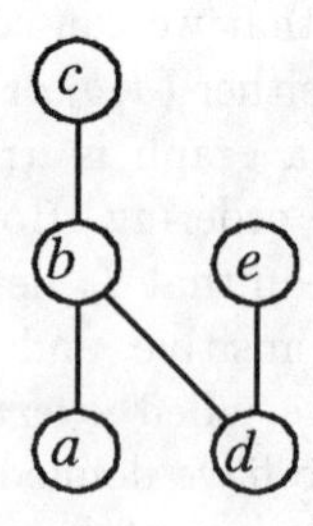

Fig. 2.

The ordered set in Figure 2 has no restriction isomorphic to $2 + 2$. We shall find an interval representation of it. The left endpoint of the interval representing a vertex should be determined by what is below that vertex, the "predecessor set" of the vertex. In Table 1, we show the predecessor sets of each vertex.

vertex x	Pred(x)	Suc(x)	PredSuc(x)
a	$\emptyset$	$\{b,c\}$	$\{a,d\}$
b	$\{a,d\}$	$\{c\}$	$\{b,a,d\}$
c	$\{b,a,d\}$	$\emptyset$	$\{a,b,c,d,e\}$
d	$\emptyset$	$\{b,c,e\}$	$\{d\}$
e	$\{d\}$	$\emptyset$	$\{a,b,c,d,e\}$

Table 1. The predecessor, successor, and predecessor-successor sets of the order in Figure 2.

Notice that the predecessor sets form the linearly ordered set

$$\emptyset \subseteq \{d\} \subseteq \{a,d\} \subset \{b,a,d\}.$$

Since the endpoints will be linearly ordered, we might try to use this set as our set of endpoints. This will do fine for the left endpoints, but it is not clear what we should do for right-hand endpoints. Clearly the right-hand endpoint of x must be determined by what is over x. However, as you see from Table 1, the family of predecessor and successor sets need not be linearly ordered, so it can't serve as the set of endpoints. Nonetheless, if we could somehow associate a predecessor set to each successor set of x, this would give us a predecessor set that might be an appropriate right endpoint of x.

We note that because the predecessor sets are linearly ordered by set inclusion, the intersection of a number of predecessor sets is again a predecessor set. Thus, the intersection of the predecessor sets of all the successors of x is itself a predecessor set, and it is the largest predecessor set that preceeds all successors of x. We denote this set by PredSuc(x); formally

$$\text{PredSuc}(x) = \bigcap\{\text{Pred}(y) | x \leq y\}.$$

Note that if x has no successors, then we have defined PredSuc(x) to be the intersection of an empty collection of subsets of X. Set theoretically, it is standard to let the intersection of the empty collection of subsets of X be X itself, and thus we define PredSuc(x) to be X when x has no successors. Then the family of predecessor-successor sets PredSuc(x) is still linearly ordered, and as we see in Table 1 for our example, Pred(x) is properly contained in PredSuc(x). We denote by $PS(X,P)$ the set of predecessor-successor sets. In Figure 3 we show schematically the linearly ordered set $PS(X,P)$ and the intervals [Pred(x), PredSuc(x)] for the ordering of Figure 2.

Note that whenever $x < y$, PredSuc(x) $\subseteq$ Pred(y). The order we get by defining $[x_1, y_1] \leq [x_2, y_2]$ when $y_1 \leq x_2$ is shown to the right of the intervals. Note that we obtain the ordered set of Figure 2 again. Thus it appears that

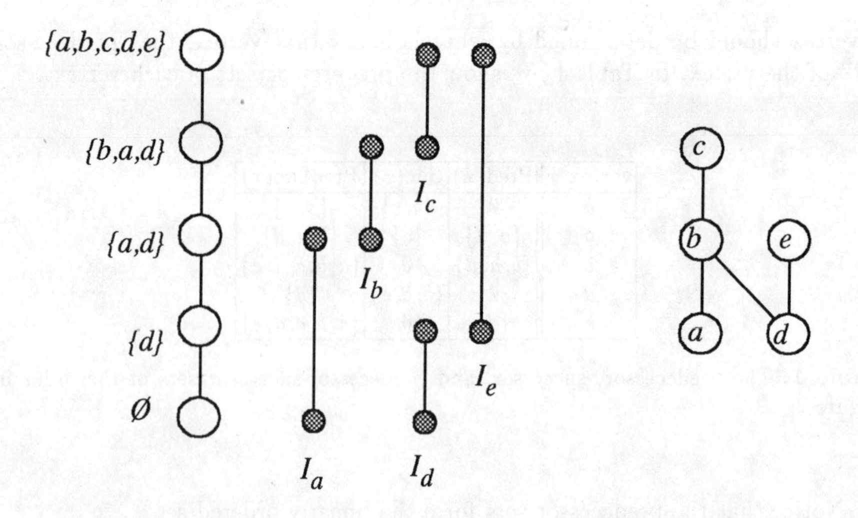

Fig. 3.

assigning each element x the interval $[\mathrm{Pred}(x), \mathrm{PredSuc}(x)]$ gives an interval representation, and our first theorem says this is always the case.

Theorem 1. *Let X be ordered by P. The mapping that takes x to the interval*

$$[Pred(x), PredSuc(x)]$$

in the predecessor-successor family of (X, P) defines an isomorphism from (X, P) onto a subset of the set of all nontrivial intervals of $PS(X, P)$, ordered by $[a, b] < [c, d]$ if and only if $b \leq c$.

The proof of Theorem 1 consists of showing that $x \leq y$ if and only if

$$\mathrm{PredSuc}(x) \subseteq \mathrm{Pred}(y);$$

this is a direct result of the definitions, so we omit the details [3]. Note that we did not hypothesize that (X, P) has no restriction isomorphic to $2+2$. However, if there is no such restriction, then $PS(X, P)$ is linearly ordered by set inclusion, giving us a basic theorem of Fishburn.

Corollary 2 (Fishburn). *The ordered set (X, P) is an interval order if and only if it has no restriction isomorphic to $2+2$.*

We may conclude immediately the corresponding result for graphs.

Corollary 3 (Gilmore and Hoffman[20], Ghoula-Houri[19]). *If G is a comparability graph that has no induced subgraph isomorphic to a cycle on four vertices, then it is an interval graph.*

Corollary 2 is less pleasing than Corollary 1 because it provides its "forbidden subgraph" characterization only in terms of cocomparability graphs. Lekkerkerker and Boland introduced the idea of an asteroidal triple to give a characterization closer to the spirit of forbidden subgraphs. They defined an asteroidal triple (AT) to be 3 nonadjacent vertices, each two connected by a path containing no neighbors of the third. In each graph in Figure 4, the three vertices shown as solid circles form an asteroidal triple.

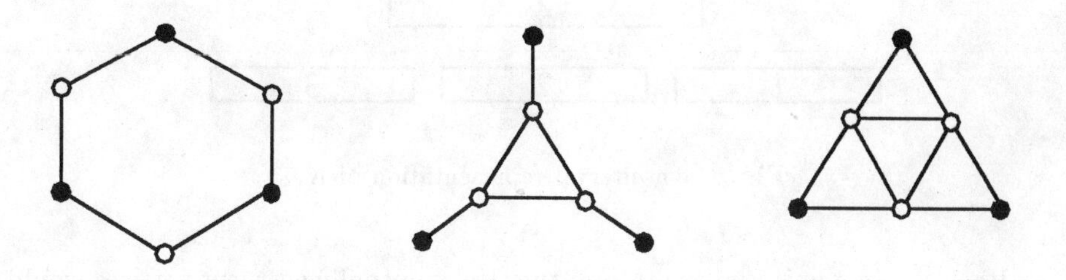

Fig. 4. Asteroidal triples

Lekkerkerker and Boland [26] proved the next two theorems.

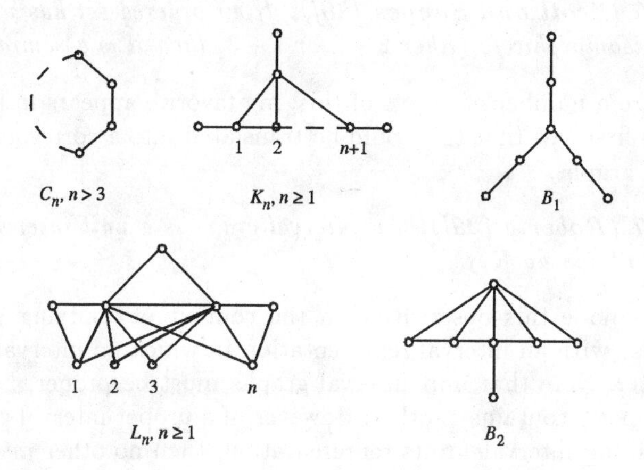

Fig. 5. The Boland-Lekkerkerker forbidden subgraphs

Theorem 4. *A graph G is an interval graph if and only if it has no induced subgraph isomorphic to a cycle on $n \geq 4$ vertices and has no asteroidal triples.*

Theorem 5. *G is an interval graph if and only if it has no induced subgraph isomorphic to one of the graphs described in Figure 5.*

An interval graph which may be represented by intervals all of the same length is called a *unit interval graph.* Correspondingly, an interval order which may be represented by intervals all of the same length is called a *unit interval order*, or, more commonly, a *semiorder.*

The complete bipartite graph on parts of size 1 and 3, K_{13}, is an interval graph (a representation is given in Figure 6).

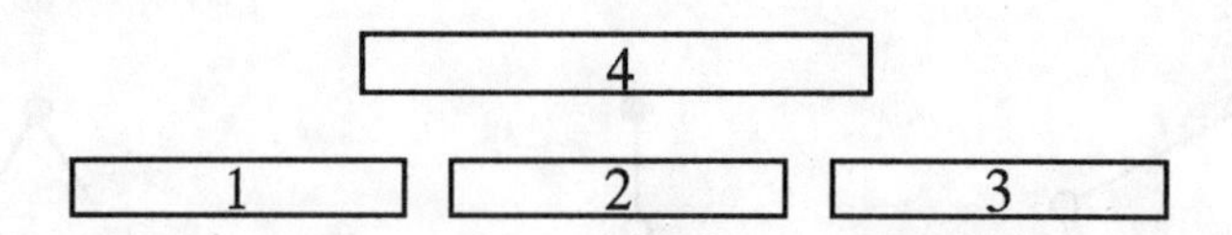

Fig. 6. An interval representation of $K_{1,3}$

However, in a unit representation, the three mutually adjacent vertices would have to be represented by three non-intersecting intervals of the same length (as in Figure 6) and a fourth interval of this length could overlap no more than two of these intervals simultaneously. Thus K_{13} is not a unit interval graph. Translated to order-theoretic terms, a semiorder can contain neither $2+2$ (to be an interval order) nor $3+1$ (by the remarks above.)

Theorem 6 (Scott and Suppes [30]). *If an ordered set has no four element restriction isomorphic to either* $2+2$ *or* $1+3$, *then it is a semiorder.*

There are a number of proofs of this; my favorite appears in [2].

Roberts first saw that this could be translated into a corresponding theorem for interval graphs.

Theorem 7 (Roberts [29]). *An interval graph is a unit interval graph if and only if it contains no* K_{13}.

Roberts made this observation in the context of studying *proper interval graphs*, those with an interval representation in which no interval properly contains another. Note that unit interval graphs must be proper, because no unit interval properly contains another. However, if a proper interval graph has three non-intersecting intervals in its representation, then no other interval can overlap them all, for it would then properly contain the interval in the middle. This gives us a second theorem.

Theorem 8 (Roberts [29]). *An inteval graph is proper if and only if it is unit; similarly an interval order is proper if and only if it is a semiorder.*

In the remainder of this paper, we consider two kinds of generalizations of the idea of an interval order. First, we need not choose our intervals in a linearly ordered set; we can define the interval $[x, y]$ in an ordered set $(X, \leq)$ by

$$[x, y] = \{z | x \leq z \leq y\}.$$

For any family of nontrivial intervals (trivial intervals have 0 or 1 points) of an ordered set $(X, \leq)$ we can define $[x_1, y_1]$ to be less than $[x_2, y_2]$ if and only if $y_1 \leq x_2$. Theorem 1 tells us that every ordered set has such a representation. Given a family $\mathcal{P}$ of orders, we say a *$\mathcal{P}$-based interval order* is an order isomorphic to the ordering of a family of nontrivial intervals chosen from an order in $\mathcal{P}$. One natural question for us to ask is for what families $\mathcal{P}$ we can obtain forbidden restriction characterizations of $\mathcal{P}$-based interval orders. Above we discussed the fact that an order is a linear-based interval order if and only if it has no restriction isomorphic to $2 + 2$.

2 Weak-based Interval Orders

In order to extend the idea of interval orders in a natural way, we must extend the family of linear orders, the family in which we choose our intervals, in a natural way. There are many natural extensions of the idea of linear orders; semiorders and interval orders are among them. Perhaps the most natural extension is the idea of a weak order. Note that a linear order may be defined as one which has no restriction isomorphic to the ordered set $1 + 1$ (Figure 7).

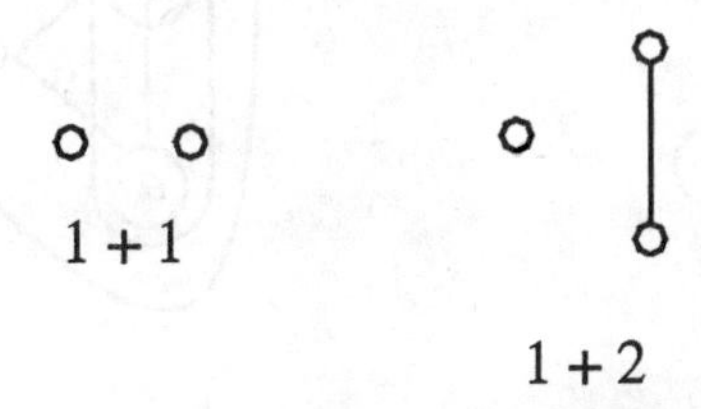

Fig. 7. The ordered sets $1 + 1$ and $1 + 2$.

A weak order may be defined as one which has no restriction isomorphic to $1 + 2$, also shown in Figure 7. For a finite set X linearly ordered by $\leq$, there is a one-to-one mapping f from X to $\{1, 2, \ldots, |X|\}$ so that $x < y$ if and only if $f(x) < f(y)$. In other words, each element of X gets a different "rank," and elements of higher rank are above elements of lower rank. For a finite set X weakly ordered by $\leq$, there is a mapping f (not necessarily one-to-one) X to $\{1, 2, \ldots, |X|\}$ so that $x < y$ if and only if $f(x) < f(y)$. Thus again, each element gets a rank, and elements of higher rank are above elements of lower rank. The only difference is that a given rank can be assigned to any number of elements.

In Figure 8 we show a six element weak order and then four intervals that represent the order $2 + 2$. There are many ways to represent the order $3 + 1$. In Figure 9 we chose the three intervals $[a, b]$, $[b, c]$, and $[c, d]$; then any interval which contains g and f will be incomparable to all three of these. We show two different choices for this interval in Figure 9, the interval $[f, g]$, and the interval $[a, h]$.

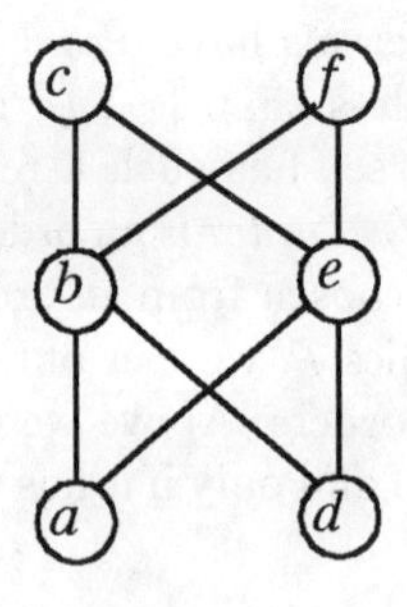

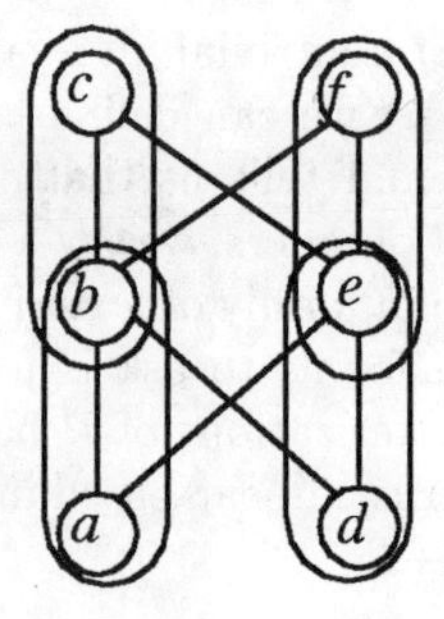

Fig. 8. A weak order and four intervals representing $2 + 2$.

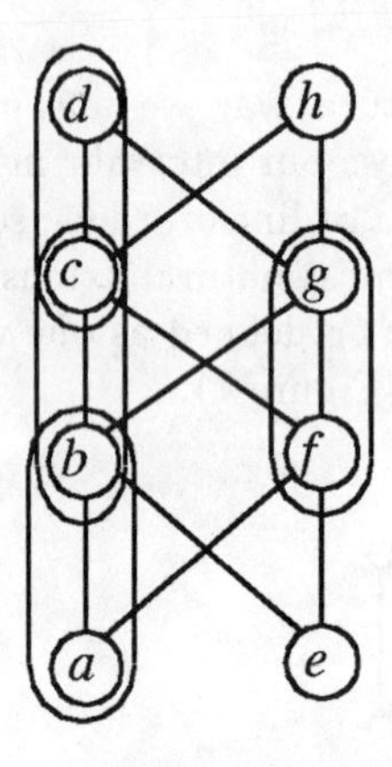

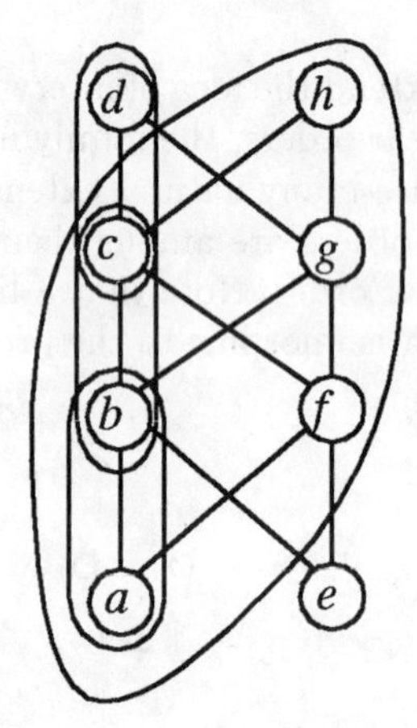

Fig. 9. Two representations of $1 + 3$.

The second choice shows us how we can obtain any ordered set of the form $1 + n$ as a weak interval order. However, it turns out that we cannot obtain the ordered set $2 + 3$ as an interval order. One argument that shows this can be obtained by staring at Figure 9 until it is clear that in order to be incomparable to all intervals of the "3," *both* intervals of the "2" would have to include both f and g and thus could not be comparable. (The fact that we used two point intervals for the "3" is irrelevant.)

It is an exercise in applying the definition of when $[x_1, y_1] < [x_2, y_2]$ to show that a weak-based interval order can contain none of the orders shown in Figure 10.

Bonin discovered the four orders shown in Figure 10, and then proved the following.

Theorem 9. *An order is a weak-based interval order if and only if it contains no restriction isomorphic to* $3 + 2$, $N + 2$, *a 6-fence, or a 6-crown.*

It is straightforward to show that no order in Figure 10 can be a restriction

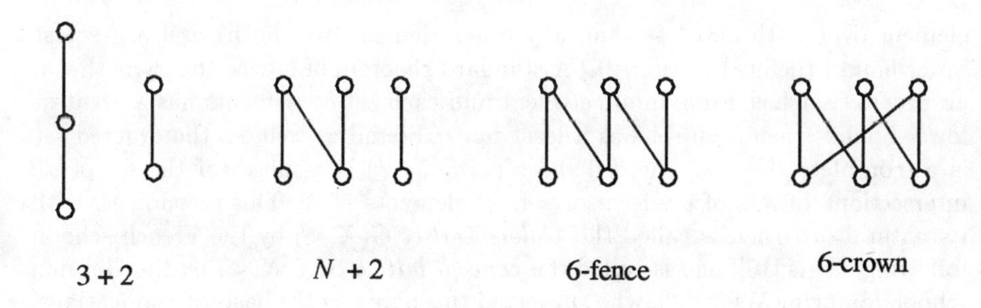

Fig. 10. Forbidden restrictions for weak-based interval orders

of a weak-based interval order; thus, once you realize you have the right family of forbidden restrictions, the main part of the proof of the theorem consists of showing the converse. However, we know from Theorem 1 that any ordered set has an interval representation in its predecessor-successor set, and so we prove that if an order has no restriction isomorphic to an order in Figure 10, then its predecessor-successor sets are weakly ordered by set-inclusion. By Theorem 1, this proves the converse. A variation on Bonin's proof which I based on an insight of Mitas's is in [4].

Notice that Theorem 3 gives a way to determine whether an order is a weak-based interval order—check all 5 and 6 element subsets to see if restricting the order to any of them gives any of the forbidden restrictions. Since the number of six element subsets of an n element set is on the order of n^6, this algorithm requires on the order of n^6 steps to recognize a weak-based interval order. In his senior thesis, Datta gave an order n^2 algorithm to solve this recognition problem and find an interval representation [11].

At this point it is natural to ask whether there is a corresponding theorem classifying a family of graphs. However, the family of graphs that would be most interesting to study as a generalization of interval graphs would have an edge between two vertices if their intervals overlap. The complements of these graphs are not necessarily transitively orientable (they include, for example, the cycle on five vertices), so "weak-based interval graphs" are not the cocomparability graphs of the weak-based interval orders. In fact the complements of the comparability graphs of weak-based interval orders appear rather contrived from a direct point of view, so we do not seem to have an interplay between interesting classes of graphs and orders. On the other hand, the "weak-based interval graphs" should be interesting to study independently of weak-based interval orders.

3 Lattice-based interval orders

Mitas [28] pointed out that the construction used in Theorem 1 actually represents the ordered set (X, P) as intervals in a lattice. (A *lattice* is an ordered set in which each pair of elements has a least upper bound (which is a unique element over both and less than any other element over both) and a greatest lower bound (defined similarly).) A standard theorem of lattice theory is that if an ordered set has a maximum element and each set of elements has a greatest lower bound, then each set has a least upper bound as well, so the ordered set is a (complete) lattice. The ordered set $PS(X, P)$ is a subset of the set of *all* intersections of sets of predecessor sets of elements of X. This second set, with its natural ordering, is called the *Galois Lattice* $G(X, <)$ by the French school, following Cogis [10], and is called the *concept lattice* $B(X, X, <)$ by the German school, following Wille [32], who suggested this name on the basis of applications. To see that it is a lattice, note that the greatest lower bound of a set of elements is its intersection, and the set X, the intersection of the empty family of predecessor sets, is a maximum element. Further, Mitas has shown that this lattice is universal in the sense that if we may represent an ordered set $P = (X, <)$ as intervals of a lattice L, then there is an isomorphism of $B(X, X, <)$ (we follow Mitas's notation) into L so that we may take the interval of L that represents x to be the interval from the image of $\text{Pred}(x)$ to the image of $\text{PredSuc}(x)$.

In the two cases we have studied so far, linear-based and weak-based interval orders, we obtain our representations as intervals in a family of ordered sets (or lattices) described by forbidden restrictions. (The examples of weak orderings that we gave were not lattices; in Figure 11 we show the lattice $B(X, X, <)$ containing the weak order in Figure 8.)

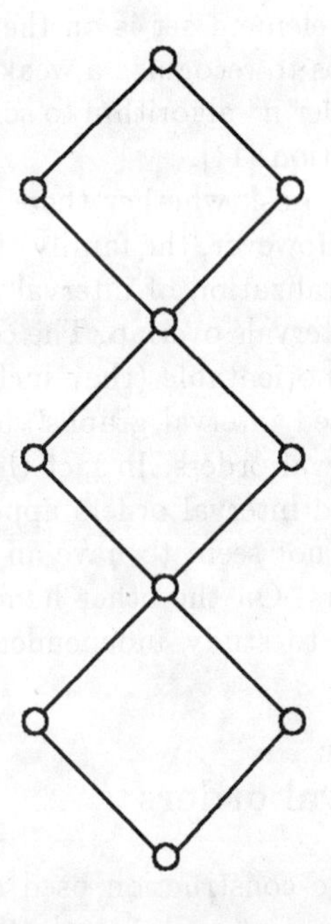

Fig. 11. The concept lattice of the weak order of Figure 8.

For linear-based interval orders, the forbidden restriction is $1+1$ and for weak orders it is $1+2$. In discussions with me, Mitas worked out an additional case in which the forbidden restriction is the ordered set N (see Figure 10), the so-called series-parallel ordered sets. This led Mitas to ask when we can represent a family of ordered sets as intervals in a family of lattices defined by a single forbidden restriction and closed under the formation of concept lattices. She has a complete description of all such lattices, a description originally worked out by Duffus and Rival [12]. They include, in addition to the linear orders and the weak-order lattices, the interval-order lattices, the series-parallel lattices, and "crown-free" lattices (see Figure 10 for a crown). Following that, she has a complete characterization of the orders that arise as interval orders from these families. The interesting ones are the series-parallel-based interval orders, and the interval-based interval orders. An order is a series-parallel-based interval order if and only if it contains no restriction isomorphic to any of the orders in Figure 12.

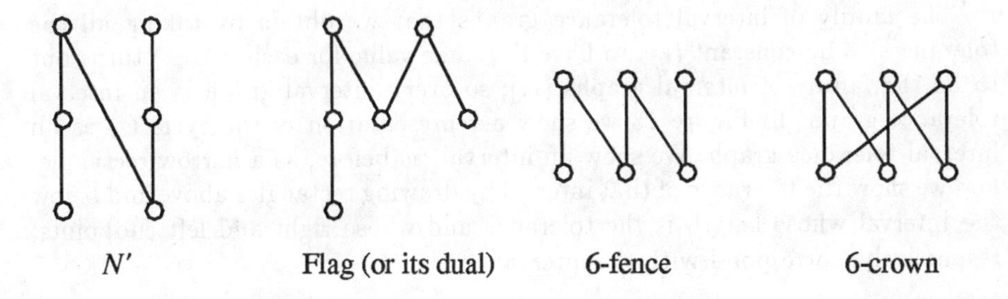

Fig. 12. Forbidden restrictions for series-parallel-based interval orders.

Further, Mitas has given a forbidden restriction characterization of interval-based interval orders that involves 37 forbidden restrictions on 6 to 8 points! Among these 37 are $3+3$ and $N+N$; in fact, in a sense all the rest of her restrictions are "between" these two.

No-one has had a chance to develop theories analogous to the theory of unit interval orders in this case, or to consider the implication of using intervals in a family of ordered sets which need not be lattices and need not be defined by forbidden restrictions. Hammer [21] has observed that all ordered sets have interval representations in Boolean algebras with $x < y$ if the right endpoint of I_x is strictly less than the left endpoint of I_y. To my knowledge, this is the only work on representing by intervals ordered by the strict order relation on appropriate endpoints, and no-one has followed up on Hammer's observation to see where it leads.

4 Tolerance Graphs and Tolerance Orders

Another natural generalization of the ideas of interval graphs and interval orders is the idea of interval tolerance graphs, introduced by Golumbic and Monma [17]. For this generalization, we continue to take our intervals on the real number line, but we modify our notion of adjacency. We say that a graph G with vertex set V and edge set E is an interval tolerance graph if we may associate an interval I_v of real numbers and a nonnegative real number t_v, called the tolerance of v, to each vertex v of G in such a way that two vertices are adjacent if the amount of overlap of the two intervals associated with them is at least as big as one of the two tolerances. In symbols, vertices u and v are adjacent if

$$|I_u \cap I_v| \geq min(t_u, t_v),$$

where $|I|$ stands for the length of the interval I. (Jacobson, McMorris and Mulder [22] have further generalized the idea by replacing the "min" in our symbolic description by an arbitrary function ϕ. Some interesting classes of graphs arise when ϕ is different from "min".)

The family of interval tolerance graphs that we obtain by taking all the tolerances to be constant (i.e. to have the same value for each vertex) turns out to be the family of interval graphs [17], so every interval graph is an interval tolerance graph. In Figure 13 we show a representation of the cycle C_4 as an interval tolerance graph. We show an interval, as before, as a narrow rectangle, and we show the tolerance of that interval by drawing rectangles above and below the interval whose length is the tolerance and whose right and left endpoints, respectively, correspond with the interval.

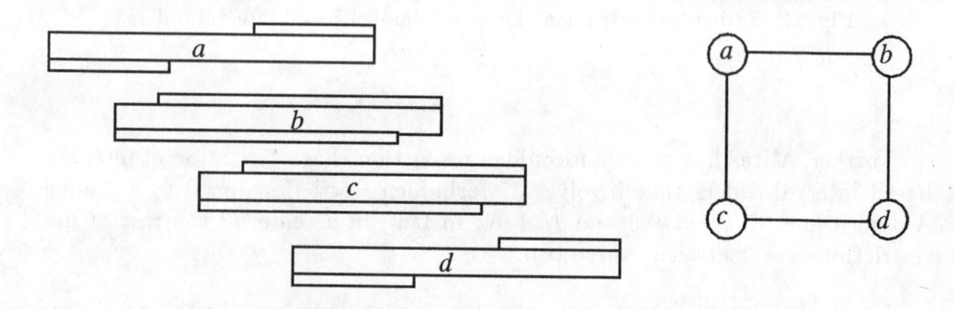

Fig. 13. A tolerance representation of a cycle on four vertices

The drawing suggests that we should also consider "bitolerance" graphs and orders; namely, the tolerance rectangles for a given interval could have different length, and two intervals are adjacent if one "contains" a tolerance rectangle of the other. We make a few remarks about bitolerance graphs and orders as we go along; except for these remarks, little is known about various classes of bitolerance graphs (though, as we point out later, the family itself is already known by another name).

The relationship between interval graphs and interval orders suggests that the complement of a tolerance interval graph should be a comparability graph. However, in Figure 14, we show a graph with a tolerance representation and then its complement, which is not a comparability graph. Experimentation with assigning directions to the edges shows that the complement does not have a transitive orientation.

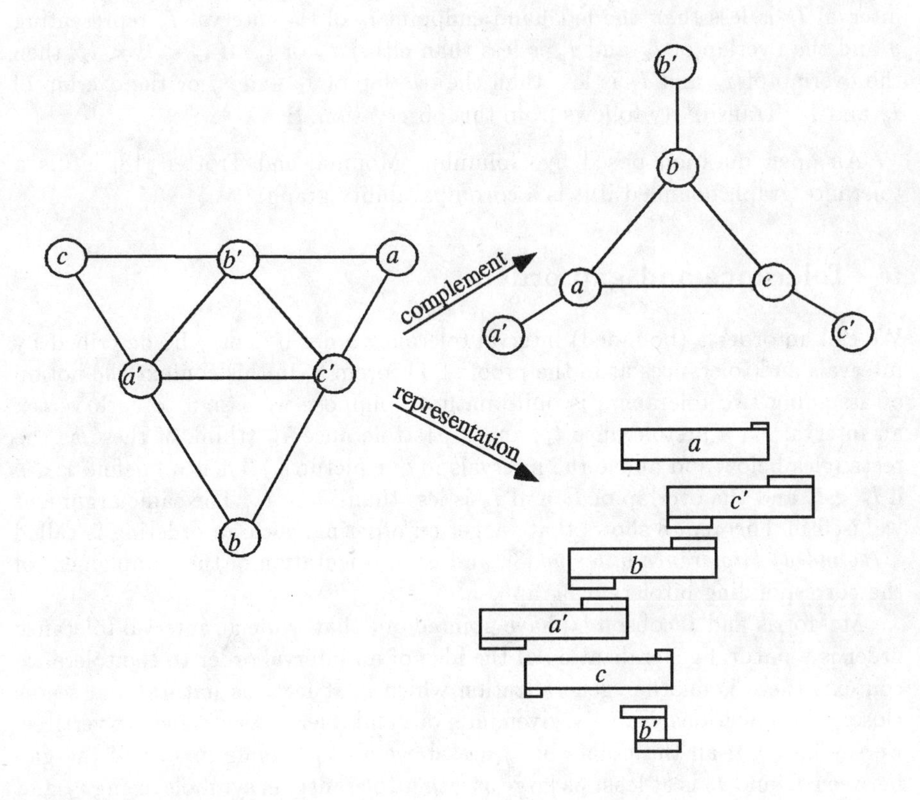

Fig. 14. A tolerance graph whose complement is not a comparability graph.

Notice that the tolerance of vertex b' in Figure 14 is larger than the length of the interval representing vertex b'. Since the overlap of the interval representing b' with any other interval is no more than the length of the interval for b', the vertex b' can only be made adjacent to another vertex because of the tolerance of that vertex, not because of the tolerance of b'. Thus we would get the same graph even if we set the tolerance of b' to be infinity. In this sense, the tolerance of b' is unbounded. We say a tolerance representation of G is *bounded* if $t_v \leq |I_v|$ for each vertex v of G and that G is a *bounded tolerance graph* if it has a bounded tolerance representation. Otherwise we say G is an *unbounded tolerance graph.* It turns out that only unbounded tolerance graphs can fail to be cocomparability graphs. For one proof see the paper by Golumbic, Monma and Trotter [18]. We give a proof that is more suggestive for what comes later.

Theorem 10. *A bounded tolerance graph is a co-comparability graph.*

Proof. Suppose G has a bounded representation. Without loss of generality, we may assume that no two endpoints of intervals of this representation are the same. Orient the complement of G by $x < y$ if the left hand endpoint, l_x, of the interval I_x is less than the left hand endpoint l_y of the interval I_y representing y and the overlap of I_x and I_y is less than either t_x or t_y. If $l_x < l_y < l_z$, then the overlap of I_x and I_z is less than the overlap of I_x and I_y or the overlap of I_y and I_z. Transitivity follows from this observation. ∎

An open question posed by Golumbic, Monma, and Trotter [18] is "Is a tolerance graph bounded if it is a cocomparability graph?"

5 Tolerance and gap orders

We call an order a (bounded) interval tolerance order if it may be described by intervals and tolerances as in the proof of Theorem 8. In this context the notion of assigning two tolerances is quite natural. Suppose we assign to each vertex an interval I_x, a pretolerance t_x, and a posttolerance T_x (think of these as the rectangles below and above the intervals in our pictures.) Then we define $x < y$ if $l_x < l_y$ and the overlap of I_x and I_y is less than T_x or t_y. The same argument we used in Theorem 8 shows that we get an ordering; such an ordering is called a *(bounded) bitolerance ordering* [8], and is an orientation of the complement of the corresponding bitolerance graph.

McMorris and Jacobson [7] have pointed out that while an interval tolerance order is a natural generalization of the idea of an interval order to the tolerance context, there is another generalization which is at least as natural and seems closer to applications. Namely, given intervals and tolerances assigned to vertices, we say $x < y$ if all the points in I_y are above all the points in I_x and the gap between I_x and I_y is at least as large as either tolerance; in symbols, using r_x and l_y for the right-hand endpoint of x and the left hand endpoint of y respectively,

$$x < y \ if \ l_y - r_x \geq max(t_x, t_y).$$

This is clearly irreflexive, and it is transitive because the "max" function satisfies the triangle inequality. An order defined in this way is called an *interval gap order* (or *max interval gap order*). Interval gap orders arise naturally in terms of scheduling. Suppose we are scheduling a meeting involving several kinds of activities, some more draining than others, so that people need a certain amount of time to "unwind" between sessions. There is no reason why we need the same amount of time before and after a given activity or why we need the same amount of time for different activities. Using pretolerances and posttolerances in the natural way, this leads us naturally to the idea of an interval (bi)gap ordering. (Once again in the case of pre- and posttolerances, one verifies that the natural relation is an ordering in the usual way.) It turns out that while gap and bigap orderings appear to be a different way of looking at things, they are special cases of tolerance and bitolerance orderings respectively. We say an interval (bi)tolerance representation of an ordering is *totally bounded* if, for each vertex, twice the tolerance (the sum of the tolerances) is less than the length of the interval representing that vertex.

Theorem 11. *An interval (bi)gap order is a totally bounded (bi)tolerance order.*

The proof of the theorem consists of replacing the interval $[l_x, r_x]$ in the gap ordering with $[l_x - t_x, r_x + T_x]$ in the tolerance ordering and using the same tolerances. We may conclude that a max gap interval order has a tolerance interval order representation in which each tolerance is at most half the interval length. This suggests that it would be interesting to study interval tolerance orders in which each tolerance is exactly 50% of the interval length, i.e. *50% tolerance orders.*

Theorem 12 (Langley). *An ordering is a max gap unit interval order if and only if it is a 50% tolerance order.*

Proof outline. Given a interval gap representation in which each interval has the same length u, replace the interval $I_x = [l_x, r_x]$ by $I'_x = [l_x - u/2 - t_x, r_x + u/2 + t_x]$. This construction is reversible. ∎

Another theorem discovered independently by Isaak and by Langley provides a tie between unit interval gap orders and unit interval tolerance orders.

Theorem 13. *An ordering is a 50% interval tolerance order if and only if it is a unit tolerance order.*

Proof outline. Given a 50% tolerance representation, choose a unit u larger than any interval length. Replace an interval with center c_x by $[c_x - u/2, c_x + u/2]$, and replace its tolerance t_x by $u - t_x$. (Note: we may assume without loss of generality that a unit interval tolerance order is bounded.) If two intervals previously overlapped by more than the half-length h_x, their replacements overlap by at least the new tolerance $u - t_y$. This construction is reversible. ∎

Since our last two theorems show that unit interval tolerance orders are identical with unit interval gap orders, it would be nice to have a direct construction that proves this. Langley [23] has a very nice transformation (which arises from the fact that interval tolerance orders are intersections of two related interval orders) that follows the construction of Theorem 9 to carry unit interval gap orders directly to unit interval tolerance orders.

Still another description of orders is in terms of threshold functions. A *threshold representation* of an ordered set (X, P) is a function c from X to the real numbers and a function T of two variables from ordered pairs of elements of X to the reals such that $x < y$ if and only if $c(x) + T(x, y) < c(y)$. We think of

$c(x)$ as a coordinate assigned to x on a real number line and thus we are saying x is below y if the coordinate for x is below that for y by at least the threshold value that T specifies for x and y. It turns out that any order may be described by a threshold representation, and the 50% tolerance orders are identical with the *max threshold orders*, those in which we assign a nonnegative real threshold T_x to each x and define $T(x, y) = max(T_x, T_y)$. [1]

In light of the equivalence of unit and proper interval orders, it is natural to ask whether proper interval gap orders are proper interval tolerance orders, and whether all the various kinds of proper and unit interval gap and tolerance orders are equivalent.

It is possible to show that Langley's construction mentioned above for unit gap orders can be carried out in such a way as to convert proper interval gap orders to proper interval tolerance orders. However, as opposed to the situation with unit intervals, it is not clear that we can carry out the construction in reverse if we start from an abitrary proper tolerance order. Langley [23] points out that if we can find a proper interval tolerance order which is not a proper interval gap order, then we have shown that proper interval tolerance graphs cannot be the same class as unit interval tolerance graphs because, by our remarks above, unit interval tolerance graphs have proper interval gap representations. Thus it would be especially nice to settle the question of whether proper interval tolerance orders and unit interval tolerance orders are the same.

This question has actually already been answered in another context. Bogart, Fishburn, Isaak, and Langley [5] have shown that the order given in Figure 15 (along with its incomparability graph) is a proper interval tolerance order, but not a unit interval tolerance order (and thus the graph is a proper but not unit interval tolerance graph). These are the first members of infinite families of proper interval tolerance orders and graphs that are irreducible in the sense that deleting any one point makes them unit as well.

The proper representation which we presently know for the order is not totally bounded (i.e. the tolerances are not half the interval lengths or less.) However, it is quite possible that the order has a totally bounded representation in which case it would be an interval gap order. Further, it is possible that it is a proper interval gap order. Both these questions remain open.

In the case of bitolerance orders, it turns out that proper orders are unit.

Theorem 14 (Bogart and Isaak [6]). *Let P be an ordering of a set X. The following are equivalent.*

(1) P is a proper bitolerance order of X.
(2) P has a linear extension with the properties that
  (a) If the restriction of P to $\{a_1, a_2, b_1, b_2\}$ is a $2+2$ with $a_1 < a_2$ and $b_1 < b_2$ in P, then in the linear estension, either $a_1 < b_1 < b_2 < a_2$ or $b_1 < a_1 < a_2 < b_2$, and
  (b) If the restriction of P to $\{a, c_1, c_2, b\}$ is an N in P with $c_1 < c_2$, $a < c_2$, $b > c_1$, then in the linear extension, $a < c_1 < c_2 < b$ never occurs.
(3) P is a unit bitolerance order of X.

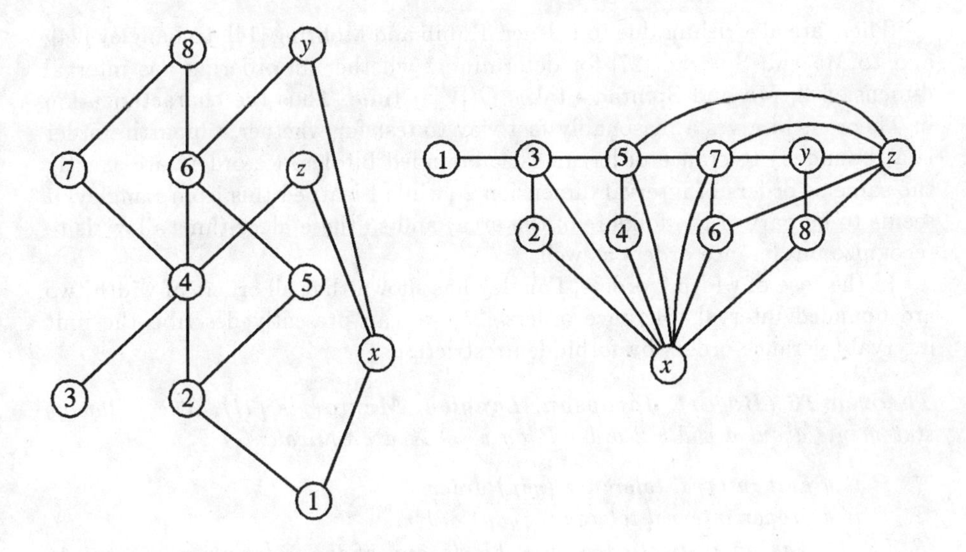

Fig. 15. A proper but not unit interval tolerance order and graph

6 Partial Characterizations of Tolerance Orders

One motivating influence in this subject is the search for characterizations of interval tolerance graphs and orders similar to those we have for interval orders. There are two special cases where we have such combinatorial characterizations of tolerance orders or unit tolerance orders, namely when the order is bipartitite (i.e. has height 1) or has width 2 (so that the tolerance graph is bipartite). These characterizations are the following two theorems.

Theorem 15 (Bogart and Trenk [8]). *For a bipartite (height 1) order, the following are equivalent.*

1. *The order is a unit (or 50%) interval tolerance order.*
2. *The order is a proper interval tolerance order.*
3. *The order is a (bounded) interval tolerance order.*
4. *The order is an intersection of two interval orders (i.e. has interval dimension 1 or 2).*

The implications $1 \Rightarrow 2 \Rightarrow 3 \Rightarrow 4$ are all standard, and so the proof of the theorem consists of showing how 4 implies 1. In addition, Bogart and Trenk showed that the four statements are equivalent for bitolerance orders as well. (Note that 4 is the same in both cases, so there is no difference between bipartite bounded tolerance orders and bipartite bounded bitolerance orders.) Felsner [13] shows that in the case of a cobipartite graph, being a tolerance graph is equivalent to being the incomparability graph of a bounded tolerance order, along with some of the equivalences of Theorem 14. Thus all cobipartite tolerance graphs are bounded.

There are algorithms due to Felsner, Habib and Mohring [14], to Langley [24], and to Ma and Spinrad [27] for determining whether an ordering has interval dimension 2; Ma and Spinrad's takes $O(|V|^2)$ time. Thus the characterization in Theorem 14 gives a reasonably fast way to test for whether a bipartite order is a (bounded) tolerance order. In fact, bounded bitolerance orders are exactly the same as orders of interval dimension 2 (while I learned this from Langley, it seems to be part of the folklore of the area) and so these algorithms allow us to recognize bitolerance orders as well.

In the case of width 2 orders, Langley has shown that all orders of width two are bounded interval tolerance orders [25] we can presently describe the unit interval tolerance orders by forbidden restrictions.

Theorem 16 (Bogart, Jacobson, Langley, Mcmorris [7]). *The following statements about a width 2 order P on a set X are equivalent.*

1. *P is a unit interval tolerance (gap) order.*
2. *P is a proper interval tolerance (gap) order.*
3. *(X, P) has no restriction isomorphic to one of the orderings in Figure 16 (or the dual of N'').*

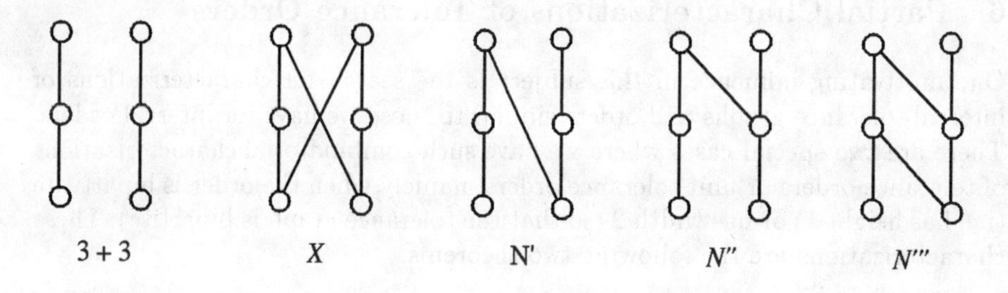

Fig. 16. Forbidden restrictions for unit interval tolerance orders of width 2.

All of the orders in Figure 16 are bounded tolerance orders, while $3+3$ is not an interval gap order, but X is. Thus we know the families of bounded interval tolerance orders and interval gap orders are different and are both larger than the corresponding "unit classes," but we do not have a characterization of the interval gap orders in the width two case.

7 Concluding Remark

By extending the family of orders in which we choose intervals beyond the linear orders or by modifying our notion of when intervals overlap enough to be adjacent (or incomparable), we convert the subjects of interval graphs and interval orders from mature, rather closed, subjects into fascinating subjects with many interesting theorems and even more interesting open questions.

References

1. K.P. Bogart. *Threshold Representations of Ordered Sets* Dartmouth College Technical report PMA-TR91-170, submitted for publication.
2. K.P. Bogart. *A Discrete Proof of the Scott-Suppes Representation Theorem for Semiorders*, Dartmouth College Technical Report PMA-TR91-173, to appear, Discrete Applied Mathematics.
3. K.P. Bogart. *An Obvious Proof of Fishburn's Interval Order Theorem*, Dartmouth College Technical Report PMA-TR91-172, Revised 1/3/92, to appear, Discrete Mathematics.
4. K.P. Bogart, Joseph Bonin, and Jutta Mitas. *Interval Orders Based on Weak Orders*, Dartmouth College Technical Report PMA-TR91-175, Revised 5/92, to appear, Discrete Applied Mathematics.
5. K.P. Bogart, Peter Fishburn, Garth Isaak, and Larry J. Langley. *Proper and Unit Tolerance Graphs*, DIMACS Technical Report 91-74, to appear, Discrete Applied Mathematics.
6. K.P. Bogart and Garth Isaak. *Proper and Unit Bitolerance Graphs and Orders* SIAM Conference on Discrete Mathematics, Albuquerque, June, 1994.
7. K.P. Bogart, M.S. Jacobson, L.J. Langley, and F.R. McMorris. *Bipartite Unit Tolerance Graphs and Width two Orders*, Dartmouth College Technical Report PMA-TR93-107, submitted for publication.
8. K.P. Bogart and Ann Trenk. *Bipartite Tolerance Orders* Dartmouth College Technical Remport PMA-TR92-101, Revised 6/92, to appear, *Discrete Mathematics.*
9. A. Bouchet. *Etude combinatoire des ordonnés finis*, Doctoral Thesis, Scientific and Medical University of Grenobel (1971)
10. O. Cogis. *Dimension Ferrers des graphes orientés* Doctoral Thesis, University of Pieree and Marie Curie, Paris (1980)
11. Sougata Datta. *The Recognition and Representation of Interval Orders and Weak Interval Orders* Senior Thesis, Dartmouth College, Hanover, NH, 1992.
12. D. Duffus and I. Rival. *Separable Subsets of a Finite Lattice, Journal of Combinatorial Theory A* **25** (1978) pp188-192.
13. Stefan Felsner. *Tolerance Graphs and Orders*, Dissertation, Technische Universität Berlin, 1992.
14. Stefan Felsner, Michel Habib, and Rolf Möhring. *The Interplay Between Ordinary Dimension and Interval Dimension*, to appear, SIAM Journal on Discrete Mathematics.
15. Peter C. Fishburn. *Interval Orders and Interval Graphs*, John Wiley (New York), 1985.
16. Martin C. Golumbic. *Algorithmic Graph Theory and Perfect Graphs*, Academic Press (New York), 1980.
17. Martin C. Golumbic and Clyde Monma. *A generalization of Interval graphs with tolerances*, Proc. 13th SE Conf. on Combinatorics, Graph Theory and Computing, *Congressus Numerantium* **35** (Utilitas Math., Winnipeg 1982), 321-331.
18. Martin C. Golumbic, Clyde Monma, and William T. Trotter, Jr. *Tolerance Graphs, Discrete Applied Mathematics* **9** (1984) 155-170.
19. A. Ghoula-Houri. *Caracterisation des graphes non orientes dont on peut orienter les aretes de maniere a obtenir le graphe d'une relation d'order C.R. Acad. Sci. Paris* **254** (1962) 1370-1371.
20. P.C. Gilmore and A.J. Hoffman. *A Characterization of Comparability Graphs and of Interval Grpahs, Canad. J. Math* **16** (1964) 539-548.

21. Peter Hammer, Personal Communication, 1991.
22. M. Jacobson, F.R. McMorris and H. Mulder. *An introduction to tolerance intersection graphs*, in *Graph Theory, Combinatorics, and Applications*, Y Alavi et. al. ed., Vol. 2 John Wiley and Sons (1991) 705-723.
23. Larry Langley, personal communication, 1992.
24. Larry Langley, *A Recognition Algorithm for Orders of Interval Dimension 2*, to appear, Discrete Applied Mathematics.
25. Larry Langley. *Orders of width 2 are Tolerance Orders* Twenty-Fifth Southeastern International Conference on Combinatorics, Graph Theory, and Computing. March, 1994.
26. C.G. Lekkerkerker and J.Ch. Boland *Representation of a Finite Graph by a Set of Intervals on the Real Line, Fundamenta Mathematica* **51** (1962) 45-64
27. T. Ma and J. Spinrad. *An $O(n^2)$ algorithm for the 2-chain cover problem and related problems., Proceedings of the Second Symposium on Discrete Algorithms*, Society for Industrial and Applied Mathematics, Philadelphia, 1991.
28. Jutta Mitas. *The Structure of Interval Orders* Thesis, Technischen Hochschule Darmstadt, 1992.
29. Fred Roberts. *Indifference Graphs*, in *Proof Techniques in Graph Theory*, F. Harary, ed. Academic Press, New York (1969).
30. Dana Scott and Patrick Suppes. *Foundational Aspects of Theories of Measurement*, J. Symbolic Logic **23** (1958), 113-128.
31. William T. Trotter. *Combinatorics and Partially Ordered Sets, Dimension Theory*, Johns Hopkins University Press, Baltimore (1992)
32. *Restructuring Lattice Theory: An Approach Based on Hierarchies of Concepts* in *Ordered Sets*, edited by I. Rival, D. Reidel, Dordrecht (1982) 445-470.

Dimension and Algorithms

J. Spinrad[1]* **

Department of Computer Science
Vanderbilt University
Nashville TN 37235, USA

Abstract. Dimension is one of the most heavily studied parameters of partial orders, and many beautiful results have been obtained. However, our knowledge of algorithms for problems on dimension is in a surprisingly primitive state. In this paper, we point to natural problems on dimension which have not been studied, and survey known results in the area.

1 Introduction

The dimension of a partial order P is the minimum number of linear extensions which yield P as their intersection. Dushnik and Miller [11] introduced dimension as a parameter of partial orders in 1941. Since that time, many theorems have been developed relating to partial order dimension, and the concept of dimension has been broadened to include intersection of sets other than linear extensions, such as interval dimension, circular-arc dimension, and others. Trotter [49] is a good source for theorems involving dimension.

In this survey, we will discuss natural algorithmic problems involving dimension. This paper takes the viewpoint that there has not been enough attention paid to algorithmic problems involving dimension, with the exception of the problem of recognizing partial orders of dimension d. The paper reviews old results, and poses new questions.

A set of linear extensions which give the partial order P as their intersection is called a *realizer* of P. A *comparability graph* is an undirected graph such that the edges can be directed to produce a transitive digraph. Comparability graphs are also called transitively orientable graphs.

2 Recognition of Partial Orders with Dimension d

The problem of determining the dimension of a partial order is one of the only algorithmic problems on dimension which has been studied carefully. When Garey and Johnson wrote the first edition of their famous book on NP-completeness [14] it was not known whether the problem of determining whether the partial

* Email: spin@vuse.vanderbilt.edu
** Project Sponsored by the National Security Agency under Grant Number R592-9632

order has dimension at most d was polynomial or NP-complete, and this was selected as one of twelve outstanding open problems in the book. The problem was solved by Yannakakis [52], in which a relatively difficult reduction from graph 3-colorability was used to show that determining whether a partial order has dimension d is NP-complete for all fixed $d \geq 3$. As we will see later, determining whether a partial order has dimension at most 2 can be done in polynomial time. One special subproblem posed by Yannakakis remains open.

Problem 1. Given a height one partial order P, is the problem of determining whether P has dimension less than or equal to 3 NP-complete?

Thus, most of the algorithmic work on the recognition problem has dealt with the case of two dimensional partial orders. Two dimensional partial orders can be recognized in $O(n^2)$ time, and this fact has been used to develop $O(n^2)$ algorithms for circular-arc graph recognition [12] and trapezoid graph recognition [31] Algorithms for recognizing two dimensional partial orders have been surveyed in earlier versions of this conference [34]. In this paper we will discuss several recent simplifications due to McConnell, which make previous asymptotically efficient algorithms much simpler.

Dushnik and Miller showed that a partial order P is two dimensional if and only if the complement of P is a comparability graph. Let P_c' be a transitive orientation of the complement of P, and P_c'' be the reversal of P_c'. If P is two dimensional, then the two linear extensions $P + P_c'$, $P + P_c''$ form a two dimensional representation of P.

Due to this characterization, recognition of two dimensional partial orders is tied closely to comparability graph recognition. Comparability graphs can be recognized fairly easily in polynomial time; the basic approach is to orient a single edge, and to find which other edges are directly 'forced' by this. All edges which have been oriented are placed on a queue, to see which other edges they force. Although details vary, the basic forcing approach has appeared a number of times in the literature [16], [15] [13]. It takes $O(n)$ time to find which other edges are directly forced by the orientation of a an edge, and this leads naturally to an algorithm with running time $O(n^3)$.

In order to present more efficient algorithms for transitive orientation, we must describe the substitution decomposition briefly. For an excellent survey of substitution decomposition, see [35].

A module (or homogeneous set) M in an undirected graph is a set of vertices such that every vertex which is not in M is either a) adjacent to every vertex of M or b) nonadjacent to every vertex of M.

Substitution decomposition decomposes a graph as follows. We start with a subgraph consisting of the entire graph. If the current subgraph is disconnected (we call such subgraphs parallel modules), decompose the subgraph into connected components. If the complement of the current subgraph is not connected (called series modules), decompose the subgraph into connected components of the complement graph. If both the subgraph and its complement are connected (neighborhood modules), decompose the subgraph into maximal proper submodules. Each subgraph of the decomposition is decomposed recursively, until all subgraphs are single vertices. The decomposition can be thought of as producing a trcc, and the decomposition tree is unique [35].

A naive approach to substitution decomposition would be rather expensive. The simplest algorithm which comes directly from the definition would involve trying all pairs of vertices, and finding the smallest module which contains these vertices; a simple version could have running time as high as $O(n^5)$. A number of clever $O(n^3)$ algorithms have appeared in the literature [2], [19].

We will discuss an incremental approach to substitution decomposition. By this, we mean that vertices are added one at a time, and the substitution decomposition of the current graph is found each time a vertex is added.

If you start with the idea of using an incremental algorithm for substitution decomposition, it is easy to develop an $O(n^3)$ algorithm. Let x be a new node which is added to the graph. The nodes of the current decomposition tree can be marked as either being adjacent to x (all descendants adjacent to x), nonadjacent to x, or split by x (at least one, but not all, descendants adjacent to x) in $O(n)$ time. From these markings, it is relatively easy to see how the decomposition tree must be changed. As a single example, let us assume that the current graph is disconnected, and x splits component C, while x is nonadjacent to the set of components $C1$. In this case, x is inserted recursively into component C. If, on the other hand, there was also a nonempty set of components $C2$ which were adjacent to x, the new decomposition would have a connected component containing x, C, and $C2$; this component has a (disconnected) submodule C_2, a submodule consisting of x, and C needs to be split using a recursive splitting procedure. Although a set of rules must be specified, I believe that any researcher who takes an incremental approach will arrive at an $O(n^3)$ algorithm.

In fact, most of the rules for adding a vertex can be computed simply by examining the markings (split, adjacent, or nonadjacent) at each node, and the time spent is proportional to the number of nodes. The only operation which makes it hard to design an $O(n)$ algorithm for adding x is when we must add x to a neighborhood module, and want to know which child (if any) x forms a module with. If we try naively to compare the adjacencies of x with the adjacencies of each submodule, the time taken is proportional to the square of the number of submodules. If the entire graph consists of one neighborhood module with no submodules (as will be true for random graphs) [37] this gives an $O(n^3)$ algorithm.

The first $O(n^2)$ incremental algorithm for substitution decomposition can be found in [40]. In that paper, the authors develop a data structure which allows you to decide which submodule may contain x in time proportional to the number of children of the module. Unfortunately, this data structure is quite complex. McConnell developed a wonderfully simple alternative approach, which will be described here.

For each pair of submodules M_1, M_2 of a neighborhood module, keep track of a vertex which is not in M_1 or M_2, and is adjacent to M_1 and nonadjacent to M_2. There must be such a vertex, or $M_1 \cup M_2$ forms a submodule, contradicting the fact that children of a neighborhood node are the maximal submodules of the node. To test which submodule the new vertex belongs in, we successively take pairs of candidate submodules, and find the vertex y which distinguishes these. By checking the adjacency of x to y, we rule out one of M_1,M_2 as a candidate module containing x, so placing x in its appropriate module takes time proportional to the number of children of M. The rules for making sets children of the same neighborhood module allow you to find distinguishing vertices between the sets in constant time, and the total number of distinguishing vertices which need to be found is $O(n^2)$. Therefore, this gives an $O(n^2)$ algorithm for finding the substitution decomposition; it should be noted that the time bound is no longer $O(n)$ per vertex added, but is $O(n^2)$ amortized over the addition of n vertices.

We now turn our attention to an $O(n^2)$ transitive orientation algorithm. If we are given the substitution decomposition of a graph G, we can perform transitive orientation during a postorder decomposition of the decomposition tree. If the next node encountered in a postorder traversal is a parallel node, there are no new edges to direct, and we continue. If the next node is a series node, choose any linear order of the children, and orient all edges between children consistently with this linear order. If the next node is a neighborhood node, create a new graph M' with one 'representative' vertex from each child of the node. If we have a transitive orientation for M', orient any edge between different children in the same direction as the edge between the representatives in the transitive orientation of M'. The only difficult step of performing transitive orientation in this way is the orientation of M', which is a neighborhood module with no nontrivial submodules.

The orientation algorithm described here has two steps; in the first step, we find a vertex which can be made a source in the transitively oriented graph, and in the second we use a technique called vertex partitioning to get a transitive orientation of the whole graph. The first step is from [33] and is much simpler than the approach in [46]; the second step is described in more detail in [46].

To find a source, we use the following procedure. Start with any vertex x, and consider a set $S = V - x$. Construct a queue, which originally contains vertex x. While there is more than one vertex in S, remove the vertex from the front of the queue; we will call this f. If f has at least one neighbor in S, remove all nonneighbors of f from S, and place them in the queue. When S contains one vertex, make this vertex a source of the directed graph. It should be clear that the source s can be found in this manner in $O(n+m)$ time. To see that this is a valid source if G is a comparability graph, assume that the edge from x is directed from s to x. Now consider the first vertex v removed from s which has an edge to s. Since v was removed from S before s, there was a vertex w removed earlier which is adjacent to s, but not to v. Since w was removed earlier, there is an edge from s to w, which implies that the edge (v,s) must be directed from s to v.

We use a technique called vertex partitioning in order to find the transitive orientation of the graph. We maintain sets of vertices, with the property that edges between vertices in different sets are directed, while edges between vertices in the same set are undirected. Originally, the sets are the source vertex s, and V - s, with all edges directed out of the source vertex.

We select a vertex x, and use x to split the sets which do not contain x into neighbors and nonneighbors of x. Consider a set S which is split into S_1, S_2, where vertices in S_1 are neighbors of x and vertices in S_2 are nonneighbors of x. We want to maintain the property that edges between different sets are directed consistently with a transitive orientation. Examine all pairs of vertices $s_1 \in S_1$, $s_2 \in S_2$. If s_1 and s_2 are adjacent, the edge must be directed from s_1 to s_2 if and only if the edge between x and s_1 is directed from s_1 to x. This can be determined in constant time. Since each vertex pair is examined only when the vertices are placed in separate subsets, the time spent directing edges is $O(n^2)$.

The last nonobvious detail for achieving the $O(n^2)$ time bound for transitive orientation is the time spent splitting sets into subsets on the basis of adjacency to x. For each vertex v, maintain a list of vertices in different sets from v which have not been split by v. Whenever a set S is split into S_1 and S_2, append all vertices of S_2 to the lists of all vertices in S_1, and vice-versa. Any vertex with a nonempty list is chosen as a splitting element, and the split is achieved by traversing the list. Each vertex enters the list of each other vertex once, so the splitting takes $O(n^2)$ time.

It is not difficult to modify the partitioning procedure to run in $O(m\log n)$ time rather than $O(n^2)$ time. This can be achieved by splitting sets during a traversal of the entire adjacency list of a vertex x, ignoring edges to vertices which are in the same set as x, if we make sure that each adjacency list is traversed $O(\log n)$ times. This can be done by maintaining a list of candidate splitting vertices, and adding v to this list if the set containing v is split into subsets such that v is in the smaller subset with respect to number of vertices in the subset.

There is a close relationship between the partitioning procedure itself and substitution decomposition. It is my belief that almost any substitution decomposition algorithm can be transformed into a transitive orientation algorithm with the same time complexity. Thus, the recent linear time algorithms for substitution decomposition [9], [33] should give linear time recognition algorithms for two dimensional partial orders.

It is important to note that the algorithm described above does not give an $O(n^2)$ algorithm for recognizing comparability graphs. The algorithm will produce a transitive orientation of a comparability graph in $O(n^2)$ time, but if the input graph is not a comparability graph, it will produce some nontransitive orientation. This does not cause a problem for recognition of two dimensional partial orders, since it is easy to verify that the pair of linear extensions produced represents the partial order correctly.

Surprisingly, the problem of verifying transitivity seems to be harder than the problem of producing a transitive orientation. This can certainly done using matrix multiplication, by checking whether the square of the adjacency matrix gives any edges which are not in G; thus, the best known asymptotic complexity for comparability graph recognition takes $O(n^{2.376})$ time [8]. However, as opposed to transitive closure, this is not known to be as hard as matrix multiplication.

Problem 2. Verification of Transitivity

Can you verify that a digraph is transitive in less time than it takes to do matrix multiplication? I would also be interested in a result of the following form: this problem is as hard as verifying the result of a Boolean matrix multiplication. Recognizing whether G is transitively reduced, i.e. G is the diagram of some poset, in less than matrix multiplication time is also open.

3 Optimization Problems

Optimization algorithms on two dimensional partial orders and their underlying undirected graphs (called permutation graphs) have been studied extensively. Results on permutation graphs are summarized in [7] while two dimensional partial orders were covered in a more general survey of computationally tractable classes of partial orders [34]. This paper will discuss optimization problems for partial orders of dimension greater than two.

Most work on optimization problems and dimension has dealt with two dimensional partial orders. If we view the problem in some ways, this is natural. If we imagine trying to solve a problem on general posets by testing to see whether the poset is in some special class, it may not be useful to work with a class which is NP-complete to recognize. Alternatively, one could argue that since recognition of d-dimensional posets is NP-complete for d greater than 2, these posets must have such complex structure that we will not be able to solve optimization problems efficiently for them.

I take another view, which makes looking for efficient optimization algorithms for NP-complete problems restricted to d-dimensional partial orders a critical open problem. We believe that dimension is a parameter which in some sense measures the complexity of a partial order; indeed, a great deal of work has been done on this parameter. If this is the case, we would expect various problems to be much easier to solve when restricted to partial orders of small dimension; the fact that such algorithms have not been found can be taken as an argument that dimension is not a good parameter for measuring complexity.

Therefore, I view a recent result of Bodlaender, Kloks, and Kratsch [3] as an important breakthrough in dimension theory. This paper shows that the parameters treewidth and pathwidth, which are NP-complete to determine for cocomparability graphs, can be computed in polynomial time for cocomparability graphs of any fixed dimension d. The approach, called the minimum separator approach, has been applied to solve a number of other problems as well [38], [10], and will be surveyed here.

Let P be a partial order of dimension d, and let C be the cocomparability graph of P. If we are given a realizer for P, we can form an intersection model for C by placing the d linear extensions of the realizer on top of each other, and joining the appearance of an element in consecutive linear extensions of the realizer by a straight line. This is essentially the intersection model for cocomparability graphs given in [17]. A scanline of the model is a set of d-1 segments running from a point along the top linear extension to the bottom linear extension; segment i is joined to segment i+1 at the i+1st linear extension.

Every minimal separator of a pair of vertices u, v of C corresponds to the set of vertices which touches a scanline. Since the number of scanlines is at most n^d, the number of minimal separators is polynomial if the dimension of the cocomparability graph is fixed.

Bodlaender, Kloks, and Kratsch showed that in an arbitrary graph, the treewidth can be computed in the following manner. To understand the idea of what is happening, we note that computing treewidth(G) corresponds to adding edges to G to get a chordal graph, so that the maximum clique in the resulting chordal graph is as small as possible. Minimal separators in chordal graphs must be cliques, so we try adding edges to make each separator S a clique, and can compute the treewidth of G with this separator as a clique from the treewidth of the pieces of G - S.

For each possible minimal separator S of a pair of vertices, calculate the maximum treewidth of $C + S$, where C is a connected component of G - S, and edges are added to make S into a clique. The minimum of the values calculated for the separators will be the treewidth of G.

In general, this does not give a polynomial algorithm. First, the number of minimal separators can be exponential. Even if the number of minimal separators is polynomial for a class of graphs, however, this does not necessarily give us a polynomial algorithm for computing treewidth for the class. Suppose that we use a recursive approach to calculate treewidth in this manner. At each stage, we may choose a polynomial number of separators to break a graph into component pieces; the number of subgraphs we have to consider can be exponential.

Bodlaender et. al. showed that if G is a cocomparability graph of dimension d, the only 'pieces' that must be computed in the recursive decomposition correspond to regions of the intersection model; that is, vertices which occur between two scanlines. This allows you to use dynamic programming to compute treewidth of a cocomparability graph of dimension d for any fixed d in polynomial time by computing treewidth of successively bigger regions of the intersection model; this is the first known problem for which small dimension is known to help solve a problem which is NP-complete if the dimension is unbounded. Since pathwidth is equal to treewidth for cocomparability graphs [36], pathwidth is also computable in polynomial time for cocomparability graphs of fixed dimension.

The result above is clearly important, but is still not completely satisfactory for one who believes that partial orders with small dimension have properties which should make algorithms simpler on these posets. The algorithm uses the intersection model to find the treewidth, and this intersection model cannot be found in polynomial time unless P = NP. Therefore, if the graph itself is given as input, we still do not have a polynomial time algorithm for calculating treewidth. Is the key to this algorithm having the representation, or is it that the class of

d-dimensional posets is in some sense well behaved?

This question was resolved by Kloks, Kratsch, and Spinrad [28]. This paper studies the treewidth problem for cocomparability graphs of partial orders of bounded dimension, but does not assume that we have a small realizer for the poset. We cannot directly use the fact that the only pieces which arise in a recursive decomposition correspond to regions of the intersection model, since we do not have an intersection model. The paper shows that all necessary subgraphs correspond to connected components defined by removal of at most two minimal separators of the original graph. Since there are a polynomial number of pairs of minimum separators, and these minimal separators can be calculated in polynomial time [27], the number of pieces which must be considered is polynomial. This allows you to calculate treewidth using dynamic programming.

In fact, the algorithm calculates the correct treewidth in polynomial time for any cocomparability graph with a polynomial number of minimum separators. Therefore, even if we do not know beforehand whether our graph is a cocomparability graph of dimension d, the algorithm is useful. We enumerate all minimal separators; if the number of minimal separators is more than n^d we stop and say that the input is in the class. If there are $O(n^d)$ minimal separators, and G is a cocomparability graph, the algorithm calculates the correct treewidth even if the dimension is large. Therefore, any result of the algorithm can be used confidently, even if we do not know the input is in the class.

This algorithm raises an interesting open question, which applies to more general models of graphs as well. The assumption is often made that any algorithm for solving an optimization problem for a special class of graphs will first create the appropriate model, and then work on the model. This makes it seem that having a good representation is the key to having a fast algorithm. However, I cannot find an example of a natural nontrivial problem on any special class of graphs or posets which is polynomial time solvable on a class of graphs or posets if the input is given in standard form, but can be solved in polynomial time if the special representation is given.

Problem 3. Find a nontrivial problem which is NP-complete if a d-dimensional realizer is not given as input, but can be solved in polynomial time if the realizer is given.

Obviously, the question of whether the dimension is less than or equal to d is an example of a trivial problem of this form.

4 Working with a realizer

In the last section, we saw that small dimension could help solve an NP-complete problem, first with the realizer given as input, and even if the realizer is not given as input. In this section, we look at natural operations on partial orders which can be solved in polynomial time. For many of these problems, a realizer of a small dimension can help solve the problem efficiently. The major questions are, can the algorithms be made to run in linear time if the realizer is given as input, and can the problems be solved more efficiently than the general case if we know that the dimension is small but no realizer is given. These seem to be natural problems, but to the authors knowledge they have never been examined before.

One reason for the study of small dimensional partial orders can be a belief that we tend to encounter in practice partial orders which arise from a dominance relation along a small number of parameters. An analagous assumption has been studied in ecology, where it was shown that certain graphs called niche-overlap graphs arising in the study of ecosystems have small boxicity in practice [45]. Semiorders, which are useful in modeling partial orders based on preference, always have small dimension [16], [43]. In some cases, such as studies of human preferences, we will not know *a priori* which parameters are giving rise to a partial order, but we hope to use the fact that the number of parameters is small to devise efficient algorithms. In that model, we would assume that the partial order itself is given as input.

Another situation arises if we know which parameters give rise to the partial order. In this case, we can get the rankings of elements along each parameter, and have a small dimensional realizer as input. Note that this does not mean we actually know the dimension of the partial order; there could be a different realizer consisting of fewer linear extensions. However, we should have methods for using a small, but not necessarily minimum, realizer efficiently; this type of problem has received very little attention. The only algorithms known to the author which use realizers are for the case that we have a two dimensional partial order; we will see that the good behavior for two dimensional partial orders can be extended to larger dimension, but some of the speed seems to be lost.

Earlier in this paper, we discussed the problem of taking a partial order as input, and constructing a realizer for the partial order. If we assume that we are given a realizer, we have essentially the reversal of this problem; given a set of d linear extensions, how long does it take to construct an adjacency list for the poset? This problem can obviously be solved in $O(dn^2)$ time, by comparing the positions of each pair of elements in each of the d lists. Note that this problem would arise if a realizer is used for storing a poset efficiently, as has been suggested as one of the uses of dimension.

The problem of constructing adjacency lists from a special representation of has been studied for various graph classes, such as visibility graphs [21]. However, constructing the adjacency list from a realizer of dimension d has only been discussed for the special case of dimension $= 2$, as part of a recognition algorithm for two dimensional partial orders [33]. For two dimensional partial orders, it is easy to construct the adjacency lists in $O(n+m)$ time, where n is the number of elements and m is the number of adjacent elements. Create a pointer from the occurrence of each element i in the first list to the occurrence of i in the second list. Step through elements in the first list sequentially; let e be the next element. Find the position of e in the second list, and step through the rest of the list, placing each element encountered on the adjacency list of e; then delete e from both lists. The elements we step through are exactly those which come after e in both lists.

No linear time algorithm seems to be known to construct adjacency lists for even three linear extensions. Nevertheless, for sparse posets with small dimension, it is possible to do better than the $O(dn^2)$ algorithm mentioned earlier. We will show that the problem can be reduced to a set of range query problems, for which efficient algorithms have been derived in computational geometry.

We maintain a set of points in d-space. These algorithms are easier to present if we call the position of an element e in a linear extension the number of elements to the right of e. Queries take the form of d dimensional boxes, and we ask what points are contained in the box. For each element, insert a point with coordinates which correspond to the position of the element in each linear extension. To determine which vertices have an edge from vertex i, create a box which runs from 0 to the position of i in each linear extension. Range queries in d dimensional space can be done in $O(log^{d-1}n$+number of elements in the range) time [42], so we can construct the adjacency list from a set of d linear extensions in $O(m+nlog^{d-1}n)$ time.

Let me state that this is not quite the best bound known for range searching. However, when using results from computational geometry, one must note that there is a general assumption that the dimension is a fixed constant, so a term of the form $O(d^k)$ would not be included in the analysis. We want to allow dimension to grow as a function of n, so one must be able to understand the algorithm to see if such a term is missing rather than simply use the result. The assumption that the dimension is constant also explains why these results based on computational geometry behave worse than naive algorithms if dimension grows at least logarithmically with the number of elements. Since we are working on a grid, it is possible that more complex techniques such as [41] can be used to reduce the time slightly, but I am not sufficently familiar with the algorithm to know whether the analysis ignores terms of the form d^k.

Our problem can also be viewed as a special case of d-dimensional rectangle containment, where each element is viewed as a d-dimensional box. To see that our problem is special, note that there is no known linear time algorithm for the general rectangle containment problem in two dimensions, whereas it is easy for us to get a linear algorithm for the two dimensional case.

Problem 4. Design a linear time algorithm for computing the poset formed by the intersection of 3 linear extensions.

The problem above, and many of the other problems in this paper, are also open for dimension greater than three. We will state only the subproblem which appears to be the most tractable.

Another problem which can be solved more quickly than the general case if we are given a small realizer is transitive reduction. Here, there does not seem to be an obvious way to use the realizer. We will again use range queries to solve this problem. Insert a set of points in d-space, where each element has a point with coordinates corresponding to the position of the point in each linear extension. To determine whether there is an edge from x to y in the transitive reduction, we first check whether there is an edge from x to y in the poset.

For each edge (x,y) of the poset, we look at the box with coordinates strictly between the positions of x and y in each linear extension; x has an edge to y in the transitive reduction if and only if this box is empty. This has a cost of m d dimensional queries, or $O(mlog^{d-1}n)$.

Here, we can do better for both the cases $k = 2$ and $k = 3$. The two dimensional case is mentioned in [29]. For each element i, it is easy to compute all neighbors of i in the transitive reduction in $O(n)$ time. Keep a value called firstneighbor, initially set to n+1. Step through elements after i in the first linear extension. Let j be the next element encountered. If the position of j in the second linear extension is less than firstneighbor, then add an edge from i to j in the transitive reduction, and set firstneighbor to the position of j.

For the special case $d = 3$, the only necessary data structure is due to van Emde Boas [50]. This data structure allows you to maintain an ordered list by inserting and deleting elements with integer values in the range 1..n and search for the predecessor of a new element in O(loglog$n$) time per operation. We will show how to use this data structure to find the transitive reduction of a partial order given in the form of three linear extensions in O(n^2loglog$n$) time, by finding edges out of each vertex in the transitive reduction in O($n$loglog$n$) time.

Let e be an element of the partial order. We will step through elements after e in the first linear extension. Maintain a set consisting of elements which are relevant to future edges of the transitive reduction out of e in the following sense; if s is in the set, there cannot be another element t which comes after e in all linear extensions, and before s in the second and third linear extensions. The elements are ordered by position in the second linear extension. Note that if there was such an element t, we would never need s to rule out a new element as having an edge from e in the transitive reduction, since if s has an edge to any element we see then t will have such an edge as well. Relevant elements appear in reverse order (from their ordering in the second linear extension) in the third linear extension.

Consider an element v encountered in our traversal of elements which occur after e in the first linear extension. We first check to see whether e has an edge to v in the transitive reduction. Examine the positions of e and v in each linear extension, to see if there is an edge in the poset. We then search for the predecessor of v among the set of relevant vertices. Let p be the predecessor of v. If p comes before v in the third linear extension, then the edges (e,p) and (p,v) imply that there is no edge from e to v in the transitive reduction. On the other hand, suppose that v precedes p in the third linear extension. Since p is the last relevant neighbor of e which precedes v in the second linear extension, p has the earliest position of these elements in the third linear extension; this implies that e has an edge to v in the transitive reduction. If there is an edge from e to v in the transitive reduction, we need to update the set of relevant vertices by adding v to the set, and deleting every vertex which comes after v in the second and third linear extensions. This can be done by stepping though the list of relevant vertices which come after p in the ordering, deleting vertices until we reach a vertex which comes before v in the third linear extension.

If v does not have an edge from e in the transitive reduction, it is clear that we spend O(loglog$n$) time to step through v. If e has an edge to v in the transitive reduction, we have some extra work involving adding and deleting elements from the set of relevant vertices. Since each vertex is added only once and deleted at most once, and the time per operation is O(loglog$n$), it takes O($n$loglog$n$) time to find all neighbors of e in the transitive reduction.

Problem 5. Design an $O(n^2)$ algorithm to find the transitive reduction of a poset given in the form of three linear extensions. Design an algorithm which is more efficient than general transitive reduction if the number of linear extensions grows at least as fast as log$n$.

Of course, transitive closure is hard to interpret in this model; it seems no different than finding the poset from the realizer. Now let us reconsider these problems, assuming that the graph is given as part of the input.

Problem 6. Is it possible to design efficient algorithms for transitive closure and transitive reduction for three dimensional partial orders, if the input is given in adjacency list form?

The algorithms described above certainly seem to use the representation heavily; one can take the viewpoint that since we cannot find the representation efficiently, they will not be of use in designing an efficient algorithm for the case where no representation is given.

However, I believe that they give evidence that performing transitive reduction should be easier on small dimensional posets than for general partial orders. Recall that for general posets, the problem is equivalent to matrix multiplication.

One approach to computing transitive reduction on d-dimensional posets is to construct a set of f(n) linear extensions which give the poset as their intersection, and to compute transitive reduction on the set as described above. There do not seem to be any good bounds known on approximating the dimension; however, it does not seem too farfetched to believe that there are easy ways to construct f(n) linear extensions for some slow growing function f(n) if the input is a 3-dimensional partial order. Of course, this is just one possible approach; it is possible that there is a set of operations which calculate transitive reduction efficiently if the input is three dimensional, and never uses the representation.

Let me finish this section by giving an outline of an algorithm for constructing the transitive closure of a two dimensional poset; details can be found in [29]. This algorithm relies on the ability to construct the representation of a 2 dimensional poset, so it is not easy to generalize to higher dimension.

This algorithm uses a straightforward divide and conquer approach. Divide the elements of the poset in two equal pieces, by taking any linear extension L and dividing it into a first half called L_1 and a second half called L_2. Compute the transitive closure of the subgraphs induced by L_1 and L_2 recursively, and construct their representations using the $O(n^2)$ recognition algorithm described earlier. We now need to add edges from vertices in L_1 to vertices in L_2. Let R_{11},

R_{12} be the realizer for the subposet induced by L_1, and R_{21}, R_{22} be the realizer for the subposet induced by elements of L_2.

We first add all edges implied by transitivity of the form (l_1,l_{21}) (l_{21},l_{22}), where $l_1 \in L_1$ and $l_{21}, l_{22} \in L_2$. We will take O(n) time to construct all such edges out of l_1. Step through R_{21} sequentially, keeping track of which neighbor of l_1 occurs first in R_{22}. Let x be the next vertex encountered. If x does not have an edge from l_1, check to see whether there is an already encountered neighbor of x which occurs before x in R_{22}; if so, add an edge from l_1 to x. If x has an edge from l_1, check to see whether x becomes the first neighbor of l_1 in R_{22}. This can be done easily in O(n) time for l_1, and thus O(n^2) time overall.

We then perform a similar process to add all edges implied by transitivity of the form $(l_{11},l_{12}),(l_{12},l_2)$. Since L_1 and L_2 are transitively closed before this process begins, the second pass will add all edges implied by transitivity. The time complexity on a graph with n vertices is governed by the recurrence T(n) = 2T(n/2) + O(n^2), which has O(n^2) behavior.

5 Dimension Approximation

This section looks at some questions and approaches to the problem of finding a small, but not minimum, number of linear extensions to realize a poset. I feel that the questions in this section are quite important, but I am not sufficiently familiar with the recent spate of developments showing that various problems cannot be approximated unless P = NP to be sure that these techniques cannot be used on the dimension problem. Very little work seems to have been done on approximate approaches to the dimension problem.

The most natural question to pose with regard to dimension is whether you can find polynomial algorithms which take a poset, and produce a set of linear extensions which represent the poset such that the ratio between the realizer found by the algorithm and the dimension is at most f(n), for some small function f(n). Yannakakis' reduction of graph 3-colorability to dimension does not directly imply that approximating dimension is as hard as approximating coloring; of course, it may be possible to modify the reduction to get this type of result.

Positive results on approximating coloring do not imply positive results for approximating dimension; algorithms for coloring 3-colorable graphs with O($n^{.5}$) or O($n^{.4}$) colors [51], [1] do not give ways to get a set of $n^{.5}$ linear extensions which represent a 3-dimensional partial order given as input. Therefore, I pose the following question.

Problem 7. Design a polynomial time algorithm which takes a partial order as input, and produces a set of at most f(n) linear extensions if P has dimension 3.

I know of no work on the problem above, so the problem may be technically open for any f(n) which is o(n). I imagine that an approximation algorithm which uses O($n^{1-\epsilon}$) linear extensions for some ϵ should be able to be developed reasonably easily.

The following question arose when I considered natural brute force approaches to recognizing 3 dimensional partial orders; it could also be used to get some very weak bounds for the problem above. If you think of trying all possiblities, one naturally thinks of trying all linear extension as the first of the three linear extensions of a realizer, and proceeding from there. However, once the first linear extension is fixed, it is unclear what we must do. If we must try all second linear extensions, the number of possibilities jumps from $n!$ to $n!^2$. Note that once the first two linear extensions are fixed, we do not need to try all possible third linear extensions. In the third linear extension, x must come before y if x has an edge to y in P or x and y are unrelated and y comes before x in the first two linear extensions. If we create a graph with these necessary relationships as edges, there is a third linear extension which completes the first two iff the graph is acyclic, and any topological sort of the graph completes the realizer.

It is not clear whether a similar approach can be used to find a pair of linear extensions which can complete a representation, once all other linear extensions are fixed. One can take the viewpoint that this looks like finding a pair of linear extensions to represent a poset, and thus 'looks like' two dimensional partial order recognition, and should be polynomial. On the other hand, if we think of the posets we are working with being the key, these are 'difficult' posets and we should expect finding linear extensions to complete them to be NP-complete. I call the problem of determining whether there are two linear extensions to complete the representation of a poset the *dimension completion* problem.

Problem 8. Is the dimension completion problem NP-complete?

The following approach came from an attempt to take a greedy approach to dimension approximation (not to be confused with greedy dimension, as studied in [4]). Consider finding a realizer by successively adding linear extensions; we can stop when for every x, y such that x does not have an edge to y, y comes before x in some linear extension. We call a nonadjacent pair x, y such that x comes before y in every linear extension an *unsatisfied pair*. A linear extension which satisfies the maximum number of unsatisfied pairs would seem to be a good linear extension to add, even though it is easy to construct examples for which adding such extensions greedily will not produce a realizer of minimum dimension.

Problem 9. Can you find a linear extension which satisfies the maximum number of pairs in polynomial time?

The reader should know before attacking this problem that adding linear extensions of this form repeatedly will give an O(log$n$) * optimal approximation algorithm for dimension, so that if the results on approximating coloring translate to dimension, we should not expect this approach to be feasible. We will show that this approach gives an O(log$n$) * optimal realizer below.

Let d be the dimension of a poset. We will show that every time we add d more linear extensions which maximize the number of unsatisfied pairs, the number of unsatisfied pairs is divided by at least a factor of two. The d linear extensions of the optimal realizer satisfy all unsatisfied pairs; therefore, if there are U unsatisfied pairs, there is always some linear extension which satisfies at least U/d unsatisfied pairs. Let U_1 be the number of unsatisfied pairs before the d linear extensions are added, and U_2 be the number of unsatisfied pairs after the extensions are added. Suppose U_2 is greater than $U_1/2$. Then the next linear extension satisfies at least $U_1/2d$ unsatisfied pairs; this implies that all previous extensions satisfied at least $U_1/2d$ unsatisfied pairs. The number of pairs satisfied by the d previous linear extensions is at least $U_1/2$, contradicting the fact that U_2 is greater than $U_1/2$.

The argument above can easily be modified to show that if there is a constant c such that you can find a linear extension which satisfies c times the maximum number of unsatisfied pairs in polynomial time, then you have an O(log$n$) times optimal approximation algorithm for the dimension problem. I believe that this is a very promising approach, especially for approximating 3 dimensional partial orders.

6 Miscellaneous Open Problems

The first open problem of this section is related to working with a two dimensional partial order, when the realizer is given as input. Suppose that you are given a pair of linear extensions which represent the poset, and you want to find the longest chain in the poset. This problem has appeared a number of times in the literature, under different names. Notice that if we renumber the elements so that the first linear extension is 1 .. n, then this is the problem of finding the longest increasing subsequence of the second linear extension.

We briefly outline the best known algorithm for this problem, since the best time bound for the problem is not well known. An O($n$log$n$) algorithm has been discovered independently several times in the literature; for a more complete description see [16], where the problem is viewed as finding the largest clique in a permutation graph. Assume elements are labeled in increasing order of their appearance in the first linear extension of the realizer. Add elements one at a time, in order of their appearance in the second linear extension. For each i there is a 'best' clique of size i among the elements which have already been added, where best means that the highest numbered element in the clique is as small as possible. We store the highest numbered element from each such clique; the highest numbered element in clique i is always smaller than the highest numbered element in clique i+1. Adding a new element e involves doing a search of the previous cliques to see which i is the last clique with highest element less than e; we then replace the value at i+1 with e, since the clique of size i can be extended by the addition of e.

Since the values stored for the cliques occur in increasing order, a simple binary search adds an element in O(log$n$) time. However, since all elements are in the range 1..n, the same operations can be done in O(loglog$n$) time per element, using van Emde Boas' data structure [50], so the problem can be solved in O($n$loglog$n$) time. The fact that this can be done in O($n$loglog$n$) time was first observed in [53]. Clearly, the longest antichain can be found in the same time bound, by simply reversing the second linear extension.

A famous theorem, which appears as a nice use of the pigeonhole principle in discrete mathematics texts such as [32] corresponds to saying that a two dimensional poset has either a chain or an antichain of size at least $n^{.5}$. The theorem is more frequently phrased in terms of increasing or decreasing subsequences of a sequence. Suppose that we want to find a chain or antichain of size at least $O(n^{.5})$. We can simply run a maximum chain algorithm, and a maximum antichain algorithm, and find such a set in $O(n\log\log n)$ time. time. However, this gets the longest chain or antichain, which is more than we actually need.

Problem 10. Find a chain or antichain of size at least $n^{.5}$ from a realizer of a 2 dimensional poset in $O(n)$ time.

This question can be extended to other classes as well. For example, perfect graphs must have either a clique or an independent set of size at least $n^{.5}$. Is it possible to find a clique or independent set of this size in a perfect graph without resorting to the very difficult algorithms for finding maximum clique and maximum independent set due to Grötschel, Lovasz, and Schrijver [18]?

Many problems on two dimensional partial orders remain open. Two of the most interesting such problems were conjectured to be polynomial by Bouchitte and Habib [5].

Problem 11. Can you count the number of linear extensions or find the jump number in a two dimensional partial order in polynomial time?

Steiner and Stewart [48] have solved the jump number problem for the special case of two dimensional height one posets.

There are no algorithms known to the author which calculate dimension on polynomial time for any interesting class of partial orders which can have dimension greater than 3. The classes of interval orders and cycle-free partial orders seem promising; indeed Bouchitte and Habib conjectured that the latter has a polynomial algorithm [5], and cycle-free partial orders have maximum dimension of four [30], [23].

Problem 12. Find a polynomial algorithm for calculating dimension on some natural class of partial orders, which can have dimension greater than 3.

The following question was posed first in [44]. Permutation graphs correspond to undirected graphs such that the edges can be directed to be the transitive closure of a two dimensional partial order. Permutation graphs have been studied extensively. We can call a graph a 2D-covering graph if the edges can be directed to be the transitive reduction of a two dimensional partial order.

Problem 13. Can 2D-covering graphs be recognized in polynomial time?

The last question deals with using dimension as an efficient storage mechanism for a class of graphs. A bipartite graph is *chordal bipartite* if there are no chordless cycles with more than four vertices. Chordal bipartite graphs have been studied extensively; [6] gives many results on chordal bipartite graphs and related classes of graphs. A recent result [47] shows that the number of chordal bipartite graphs on n vertices is $2^{cnlog^2 n}$, and gives a method of storing chordal bipartite graphs which uses $O(nlog^2 n)$ space. This is asymptotically optimal as far as space is concerned. However, determining adjacency from this space efficient representation is expensive, in fact taking $O(n^2)$ time to determine whether x and y are adjacent. Dimension might be useful in devising a mechanism which allows you to store the graphs efficiently, and also test adjacency efficiently. Consider the poset formed from a chordal bipartite graph by directing all edges from one color class to the other color class. If the dimension is $O(\log n)$, the linear extensions would be a space optimal form of storage, and would allow testing the adjacency of a pair of vertices in $O(\log n)$ time. The problem is also related to certain conjectures involving local structure [39] or implicit representation [22] of graphs.

Problem 14. Let G be a chordal bipartite graph with n vertices. If all edges are directed from one color class to the other, is the dimension of the resulting partial order $O(\log n)$?

References

1. Blum, A., An $O(n^4)$ approximation algorithm for 3-coloring (and improved approximation algorithms for k-coloring), *Proceedings of the 21st Annual ACM Symposium on Theory of Computation*, 1991, pp. 535-542.
2. Buer, H. and R.H. Möhring, A fast algorithm for the decomposition of graphs and posets, *Mat. Operations research* **8**, 1983, pp. 170-184.
3. Bodlaender, H., T. Kloks and D. Kratsch, Treewidth and pathwidth of permutation graphs, *Proceedings of the 20th International Colloquium on Automata, Languages and Programming*, pp. 114–125, Springer Verlag, Lecture Notes in Computer Science, vol. 700, 1993.
4. Bouchitte, V. and M. Habib, On the greedy dimension of a partial order, *Order* **1**, 1985, pp. 219-224.
5. Bouchitte, V. and M. Habib, The calculation of invariants for ordered sets, *Algorithms and Order*, I. Rival (ed.), NATO Series C - Vol 255, Kluwer Academic Publishers, 1989, pp. 231-279.
6. Brandstädt, A., Special graph classes — a survey, Schriftenreihe des Fachbereichs Mathematik, SM-DU-199, Universität-Gesamthochschule Duisburg, 1991.
7. Brandstädt, A., On improved time bounds for permutation graph algorithms, *Proceedings of the 18th International Workshop on Graph-Theoretic Concepts in Computer science*, 1992.
8. Coppersmith, D. and S. Winograd, Matrix multiplication via arithmetic progressions, *Proceedings of the 19th Annual IEEE Symposium on the Foundations of Computer Science*, 1987, pp. 1-6.

9. Cournier, A. and M. Habib, A new linear algorithm for modular decomposition, Research Report 94033, LIRMM, Universite Montpellier 1994.
10. Deogun, J. S., T. Kloks, D. Kratsch and H. Müller, On vertex ranking for permutation and other graphs, Computing Science Notes 93/30, Eindhoven University of Technology, Eindhoven, The Netherlands, 1993. To appear in: *Proceedings of the 11th Annual Symposium on Theoretical Aspects of Computer Science*, 1994.
11. Dushnik, B. and E. Miller, Partially ordered sets, *American Journal of Mathematics* **63**, (1941), pp. 600–610.
12. Eschen, E. and J. Spinrad, An $O(n^2)$ algorithm for circular-arc graph recognition, *Proceedings of the 4th Annual ACM-SIAM Symposium on Discrete Algorithms*, 1993, pp. 128-137.
13. Even, S., A. Pnueli, A. Lempel, Permutation graphs and transitive graphs, *Journal of the ACM* **19**, 1972, pp. 400-410.
14. Garey, M.R. and D.S. Johnson, *Computers and Intractability: A Guide to the Theory of NP-Completeness*, W.H. Freeman, San Francisco, 1979.
15. Gilmore, P.C. and A.J. Hoffman, A characterization of comparability graphs and interval graphs, *Canad. J. Math.* **16**, 1964, pp. 539–548.
16. Golumbic, M. C., *Algorithmic Graph Theory and Perfect Graphs*, Academic Press, New York, 1980.
17. Golumbic, M. C., D. Rotem, J. Urrutia, Comparability graphs and intersection graphs, *Discrete Mathematics* **43**, 1983, pp. 37–46.
18. Grötschel, M., L. Lovasz, A. Schrijver, The ellipsoid method and its consequences in combinatorial optimization, *Combinatorica* **1**, 1981, 169-197.
19. Habib, M. and M.C. Maurer, On the X-Join decomposition for undirected graphs, *Appl. Disc. Math* **3**, 1979, pp. 198-207.
20. Habib, M. and R. H. Möhring, Treewidth of cocomparability graphs and a new order-theoretic parameter, Technical Report 336/1992, Technische Universität Berlin, 1992.
21. Hershberger, J., Finding the visibility graph of a polygon in time proportional to its size, *Proceedings of the 3d Annual Symposium on Computational Geometry*, 1987, pp. 11-20.
22. Kannan, S., M. Naor, S. Rudich, Implicit representation of graphs, *SIAM Journal on discrete Math* **5**, 1992, pp. 596-603.
23. Kierstead, H. and W.T. Trotter, The dimension of a cycle-free partial order, *Order* **9**, 1992, 103-110.
24. Kloks, T., *Treewidth*, Ph.D. Thesis, Utrecht University, The Netherlands, 1993.
25. Kloks, T., H. Bodlaender, H. Müller and D. Kratsch, Computing treewidth and minimum fill-in: all you need are the minimal separators, *Proceedings of the First Annual European Symposium on Algorithms*, pp. 260–271, Springer Verlag, Lecture Notes in Computer Science, Vol. 726, 1993.
26. Kloks, T. and D. Kratsch, Treewidth of chordal bipartite graphs, *Proceedings of the 10th Annual Symposium on Theoretical Aspects of Computer Science*, pp. 80–89, Springer-Verlag, Lecture Notes in Computer Science, Vol. 665, 1993. .
27. Kloks, T. and D. Kratsch, Finding all minimal separators of a graph, Computing Science Note, 93/27, Eindhoven University of Technology, Eindhoven, The Netherlands, (1993). To appear in: *Proceedings of the 11th Annual Symposium on Theoretical Aspects of Computer Science*, 1994.
28. Kloks, T., D. Kratsch, J. Spinrad, Treewidth and pathwidth of cocomparability graphs of bounded dimension, submitted for publication, 1994.

29. Ma, T.-H. and J. Spinrad, Transitive closure for restricted classes of partial orders, *Order* **8**, 1991, pp. 175-183.
30. Ma, T.-H. and J. Spinrad, Cycle-free partial orders and chordal comparability graphs, *Order* **8**, 1991, pp. 49-61.
31. Ma, T.-H. and J. Spinrad, An $O(n^2)$ algorithm for two-chain cover and related problems, *Proceedings of the 2d Annual ACM-SIAM Symposium on Discrete Algorithms*, 1991, pp. 363-372.
32. Maurer, S. and A. Ralston, *Discrete Algorithmic Mathematics*, Addison-Wesley, Reading MA, 1991.
33. McConnell, R. and J. Spinrad, Linear-time modular decomposition and efficient transitive orientation of comparability graphs, *Proceedings of the 5th Annual ACM-SIAM Symposium on discrete Algorithms*, 1994, pp. 536-545
34. Möhring, R. H., Computationally Tractable Classes of Ordered Sets, *Algorithms and Order*, I. Rival (ed.), NATO Series C - Vol 255, Kluwer Academic Publishers, 1989, pp.105–193.
35. Möhring, R. H. and F.J. Radermacher, Substitution decomposition for discrete structures and connections with combinatorial optimization, *Annals of Discrete Mathematics* **19**, 1984, pp. 257-356
36. Möhring, R. H., Triangulating graphs without asteroidal triples, manuscript, TU Berlin, November 1993.
37. Möhring, R.H., On the Distribution of Locally undecomposable relations and independent systems, *Methods Oper. Research* **42**, 1981, pp. 33-48.
38. Müller, H., Recognizing interval digraphs and bi-interval graphs in polynomial time, FSU Jena, Germany, manuscript, December 1993.
39. Muller, J., Local Structure of Graphs, Ph.D. Thesis, Georgia Institute of Technology, 1988.
40. Muller, J.H. and J.P. Spinrad, Incremental modular decomposition, *Journal of the ACM* **36**, 1989, pp. 1-19.
41. Overmars, M., Efficient data structures for range searching on a grid, *Journal of Algorithms* **9**, 1988, pp. 254-275.
42. Preparata, F. and M. Shamos, *Computational Geometry, an Introduction*, Springer-Verlag, New York, 1985.
43. Rabinovitch, I., The dimension of semiorders, *Journal of Combinatorial Theory A* **25** 1978, pp. 50-61.
44. , Rival, I. (ed.), *Algorithms and Order*, NATO Series C - Vol 255, Kluwer Academic Publishers, 1989.
45. Roberts, F.S., *Graph Theory and its Applications to Problems of Society*, SIAM, Philadelphia, 1978.
46. Spinrad, J.P., On comparability and permutation Graphs, *SIAM Journal on Computing* **14**, 1985, pp.658-670.
47. Spinrad, J., Nonredundant Ones and Gamma-free Matrices, *SIAM Journal on Discrete Math*, to appear.
48. Steiner, G. and L. Stewart, A linear time algorithm to find the jump number of 2-dimensional bipartite partial orders, *Order* **3**, 1987, pp. 359-367.
49. Trotter, W. T., *Combinatorics and Partially Ordered Sets: Dimension Theory*, The John Hopkins University Press, Baltimore, Maryland, 1992.
50. van Emde Boas, P., Preserving order in a forest in less than logarithmic time, *Proceedings of the 16th Annual Symposium on Foundations of Computer Science*, 1975, pp. 75-84.

51. Wigderson, A., Improving the performance guarantee for approximate graph coloring, *Journal of the ACM* **30**, 1983, pp. 729-735
52. Yannakakis, M., The complexity of the partial order dimension problem, *SIAM J. Alg. Discrete Methods* **3**, 1982, pp. 351–358.
53. Yu, M, L. Tseng, S. Chang, Sequential and parallel algorithms for the maximum-weight independent set problem on permutation graphs, *Information Processing Letters* **46**, 1993, 7-11.

Upward Drawings to Fit Surfaces

S. Mehdi Hashemi[1] and Ivan Rival*[2]

[1] Department of Mathematics, University of Ottawa
Ottawa K1N 6N5, Canada
hashemi@csi.uottawa.ca

[2] Department of Computer Science, University of Ottawa
Ottawa K1N 6N5, Canada
rival@csi.uottawa.ca

Abstract. Our aim is to construct ordered sets and *upward drawings* of them to fit smooth two-dimensional surfaces using piecewise linear two-dimensional ones.

Theorem. *For any smooth two-dimensional surface S of genus g there is an ordered set P such that*

(i) P has an upward drawing, without crossing edges, on S,

(ii) P contains the ordered set crit(S) of critical points of S,

(iii) if S' is any two-dimensional surface of genus g on which P has an upward drawing, without crossing edges, then crit(S) $\subseteq$ crit(S').

Much like the approximation of an arbitrary smooth function by an interpolating polynomial, we propose to approximate smooth two-dimensional surfaces by polyhedral surfaces, that is, piecewise linear two-dimensional surfaces, themselves modelled by ordered sets and their *upward drawings*. The analogy is fairly accurate for, much like numerical interpolation techniques use the function's values and derivatives at certain points, our starting point is the set of critical points of the surface. In this respect it is natural, for our purposes, to assume that a (*smooth*) *surface* S is a

closed, compact, two-dimensional orientable manifold,
smooth enough to have a continuously turning tangent plane,
embedded in $\mathbf{R}^3$ above the horizontal plane $z = 0$,
such that every point with tangent plane parallel to $z = 0$ has a neighbourhood in which it is the only critical point.

Of course, the study of these surfaces and their critical points is not new (cf. [**Milnor** (1963)], although many new ideas are still launched (cf. [**de Rezende** and **Franzosa** (1993)]. Indeed, the common ground between flows and topology dates at least to Poincaré, whose well-known *Index Theorem*, relating the Euler characteristic of the surface to the sum of the indices of its critical points (with respect to some gradient flow), has surprising discrete analogues (cf. [**Glass**

* Supported in part by N.S.E.R.C.

(1973)]). Thus, for a cartographic map, say of countries drawn (as simply connected regions) on the surface of a sphere, the alternating sum of the number of vertices, number of edges, and the number of faces, is a constant (two) — its Euler characteristic. What is new, however, is the study of order types and their upward drawings, whose "critical point characteristics" are unlabelled, in order to interpolate a surface.

The application metaphors of this subject are diverse, too. Computational graphics, for instance, is useful in cartography as well as in dynamical systems. The graphic search of topographic maps for peaks, pits, and valleys aids in the identification, for example, of local watersheds. A qualitative description of its critical points will also determine the phase portrait of a system of differential equations. A common feature of both applications is a surface in R^3 — perhaps a sphere or a torus, or any other two-dimensional surface equipped with a smooth vector field. And, if we fix such a surface on which every critical point is isolated, then the set of its critical points is an ordered set: $x < y$ if, with respect to the z-axis, there is a strictly monotonic path on this surface from x to y.

Our aim is to use *upward drawings* of ordered sets to produce piecewise linear two-dimensional surfaces to fit smooth two-dimensional surfaces. It is customary and convenient to render an ordered set by an *upward drawing* according to which the elements of the ordered set P are drawn on a surface, traditionally a plane, as disjoint small circles, arranged in such a way that, for $a, b \in P$, the circle corresponding to a is higher than the circle corresponding to b whenever $a > b$ and an arc, monotonic with respect to a fixed direction, usually south to north, is drawn to join them just if *a covers b* (that is, for each $x \in P$, $a > x \geq b$ implies $x = b$). We say that a is an *upper cover* of b or b is a *lower cover* of a, and write $a \succ b$ or $b \prec a$. These arcs are drawn, of course, to avoid the incidence of any other circle on it (to avoid unwanted comparabilities) and, moreover, when possible, to avoid intersections, too (except where two arcs meet at a circle). The *covering graph cover*(P) of an ordered set P is the graph whose vertices are the elements of P, and with edges $x \sim y$, if either $x \succ y$ or $x \prec y$ in P.

The question, is there an effective procedure to decide of an ordered set whether it has an upward drawing on the plane, without any crossing of edges — a *planar* upward drawing — is a difficult one. Although such a *planar* ordered set must, of course, have a planar covering graph, an ordered set with a planar covering graph need not be a planar ordered set (for example, the ordered set of all subsets of a three-element set — the three-dimensional cube). Quite recently [**Garg** and **Tamassia** (1994)] have announced a solution, according to which, this decision problem is *NP-complete*, confirming that planarity-testing for ordered sets is considerably more difficult than it is for (undirected) graphs [**Hopcroft** and **Tarjan** 1974].

We are led ineluctably to study upward drawings on oriented surfaces in $\mathbf{R}^3$, whose edges are strictly monotonic paths (south to north) with respect to the z-axis.

The upward drawing of an ordered set is far from unique. In fact, there may be many associated cell complexes on the surface on which the ordered set is (cellularly) embedded. Nonetheless, there is one fairly obvious upward drawing associated with a "lifting" procedure [**Ewacha, Li,** and **Rival** (1991)].

Starting with $cover(P)$ which, as a graph, has *genus* (g say), embed $cover(P)$, without crossing edges, on an oriented two-dimensional surface (a sphere with g handles attached). It is convenient to represent this embedding inside a polygon in the $z = 0$ plane of R^3 with $2g + 2$ sides, in which $cover(P)$ is drawn planar, possibly with repeated edges and vertices, and in which g pairs of sides are tagged for identification.

Next, define any (real-valued) height function h on the vertices [$x < y$ implies $h(x) < h(y)$] and continue it by linearity onto the polygon, using a triangulation of the embedding on the polygon, thus producing a piecewise-linear homotopic image which can then be piecewise linearly reconstructed, monotonically, with respect to h in R^3, by gluing pairs of sides of the polygon (already tagged for identification).

Unlike the planar case, a triangulation for graphs embedded on two-dimensional surfaces with *genus* > 0, may be impossible — without additional points.

Although the piecewise linear two-dimensional surface, manufactured by the lifting and subsequent gluing of polygon sides, preserves the *genus* of the initial two-dimensional surface, we may have self-intersections — an *immersion* — despite the fact that the surface is topologically equivalent to one without intersections, that is, a sphere with attached handles. Unlike (covering) graphs, it is not yet known whether an ordered set (of *order genus* g) has an upward drawing on a two-dimensional surface of genus g, which itself has *no* self-intersections — an *embedding* — [**Ewacha, Li,** and **Rival** (1991)]. *The decision problem whether the upward drawing of an ordered set can be drawn on a two-dimensional surface with fixed order genus and with a prescribed list of critical points, belongs to NP* (cf. [**Musin, Rival,** and **Tarasov** (1993)]).

We shall suppose that every critical point on a two-dimensional surface is isolated, that is, each critical point is contained in a neighbourhood in which it is the only critical point. Fix such a surface S and let $crit(M)$ stand for the set of all its critical points. Here is our objective.

Theorem 1. *For any smooth two-dimensional surface S of genus g there is an ordered set P such that*

(i) P has an upward drawing, without crossing edges, on S,
(ii) P contains the ordered set $crit(S)$ of critical points of S,
(iii) if S' is any two-dimensional surface of genus g on which P has an upward drawing, without crossing edges, then $crit(S) \subseteq crit(S')$.

We say that the upward drawing of P *fits* the two-dimensional surface S, and that P interpolates S. In certain cases, (e.g. smooth two-dimensional surfaces S of genus zero), we can efficiently construct such interpolating ordered sets to fit S.

The matter is not simple — apart from the easiest case, the spherical ball S^2 with precisely two critical points, one maximum and one minimum. For instance, *there is no smooth torus with precisely three critical points, one maximum, one minimum and one saddle point, although there is a two-dimensional piecewise linear surface with precisely three critical points* (cf. [**Banchoff** and **Takens** (1975)]).

Index of Critical Point

What is required is the *index of a critical point* for directed graphs on the surface, a problem considered by several authors [**Banchoff** (1967)], [**Glass** (1973)], [**Musin** (1992)], [**Musin, Rival,** and **Tarasov** (1993)]), yet apparently, little known. Thus, suppose a directed graph is embedded without edge crossings on a closed compact surface S. Assume, moreover, that the embedding is *cellular*, that is, it partitions the surface into simply connected (null-homotopic) faces F, bounded by edges E, connecting its vertices V. Accordingly, the *Euler relation* gives $|V| - |E| + |F| = \chi(M)$. Now [**Glass** (1973)] defines *index of a critical point* for a directed graph, and establishes a discrete analogue of the *Index Theorem*, as follows. Call a pair of edges of the graph embedding *adjacent* if they are incident to the same vertex and bound a common face. A vertex of degree precisely two has precisely t! wo such adjacent pairs, while, by definition, we say that a vertex of degree one has one such adjacent pair. Next, call an adjacent pair *black* if both of its edges are directed toward the incident vertex or else both are directed away from the incident vertex. Otherwise, call the pair *white*. The *reversal number* $reverse(v)$ of a vertex v is the (even) number of its white adjacent pairs; the *reversal number* $reverse(C)$ of a face C is the (even) number of black bounding adajacent pairs of edges. Define the *index* by

$$index(v) = 1 - \frac{1}{2} reverse(v)$$

and

$$index(C) = 1 - \frac{1}{2} reverse(C)$$

Then the sum of the reversals of the vertices and of the edges equals $2|E|$, so, in view of the Euler formula,

Index Theorem *For any directed graph embedded on a compact surface S*

$$\sum_{v \in V} index(v) + \sum_{C \in F} index(C) = \chi(M)$$

Consider an acyclic triangulation of a directed graph, for instance, the directed comparability graph of an ordered set. As any face C is a triangle with precisely two black adjacent pairs, then $index(C) = 0$ and it suffices to sum the index just over all vertices. Then, calling a vertex v a "critical point" just if $index(v) \neq 0$, this equation reduces to its continuous forebear the *(Poincaré) Index Theorem*.

The upward drawing is, of course, triangle-free, so the "face" term, too, may contribute a non-zero value. We examine this case of the upward drawing in a slightly different manner. To this end, notice that, for each vertex, the incident edges on the surface "fan out" forming a circle. Mark such an incident edge + if its other endpoint u satisfies $u \succ v$ and mark it – if $u \prec v$, with respect to the order. Let n_+ stand for the number of connected components, each consisting of consecutive +'s, and let n_- stand for the number of components of –'s. There are, in effect, three cases: if

$$n_+ + n_- = 2 \text{ then } index(v) = 0$$

and the corresponding vertex is *ordinary*; if

$$n_+ = 0 \text{ or } n_- = 0 \text{ then } index(v) = 1$$

and the corresponding vertex is an *ordinary critical point*, an *extremum*—either a *local maximum* or a *local minimum*; finally,

$$n_+ + n_- \geq 2 \text{ and then } index(v) = 1 - n_-$$

in which case v is a *saddlepoint*. To compute the index of a face we appeal to its local minima. Thus, call a vertex v of the face C a *local minimum* if, with respect to the order, v is less than its (two) neighbours. And, call the face C *ordinary* if it has only one (local) minimum, else call it *nonordinary* and, in any case, set

$$index(C) = 1 - k$$

where k stands for the number of local minima of C.

In rough outline here is the algorithm to construct an upward drawing to fit a prescribed two-dimensional surface S of genus zero. Start with the ordered set $crit(M)$ of critical points of S. Draw its covering graph on the plane. Associate with each critical point its index and, starting with the saddle points, successively add vertices to the neighbourhoods of their incident edges so they acquire the required index value. Join each new vertex to a maximal or minimal element, depending whether it is constructed as lower cover, or upper cover, of a saddle point. Triangulate the resultant graph, without altering the character of the critical points. Finally, add a subdivision point to each edge of the "triangulated order" that, itself, follows from the order's transitivity.

References

T. F. Banchoff (1967) Critical points and curvature for embedded polyhedra, *J. Diff. Geom.* **1**, 245 – 256.

T. F. Banchoff and **F. Takens** (1975) Height functions with three critical points, *Illinois J. Mathematics* **76**, 325 – 335.

K. A. de Rezende and **R. D. Franzosa** (1993) Lyapunov graphs and flows on surfaces, *Trans. Amer. Math. Soc.* **340**, 767 – 784.

K. Ewacha, W. Li, and **I. Rival** (1991) Order, genus and diagram invariance, *ORDER*, **8**, 107 – 113.

A. Garg and **R. Tamassia** (1994) On the computational complexity of upward and rectilinear planarity testing, *CS-94-10*, Brown University.

L. Glass (1973) A combinatorial analog of the Poincaré index theorem, *J. Combin. Theory* **15**, 264 – 268.

J. Hopcroft and **R. E. Tarjan** (1974) Efficient planarity testing, *J. Assoc. Comput. Machin.* **21**, 549 – 568.

M. D. Hutton and **A. Lubiw** (1991) Upward planar drawing of single source acyclic digraphs, *Proc. 2nd A.C.M./S.I.A.M. Symposium on Discrete Mathematics*, pp.203 – 211.

J. Milnor (1963) Morse Theory, *Ann. Math. Stud.* **51**, Princeton Univ. Press.

O. R. Musin (1992) On some problems of computational geometry and topology, *Lect. Notes Math.* **1520**, 57 – 80.

O. R. Musin, I. Rival and **S. Tarasov** (1993) Upward drawings on surfaces and the index of a critical point, *Journal Combinatorial Theory A*, to appear.

A Cleanup on Transitive Orientation

Klaus Simon Paul Trunz

Institut für Theoretische Informatik
ETH-Zentrum
CH-8092 Zürich

Abstract. In the past, different authors developed distinct approaches to the problem of transitive orientation. This also resulted in different ideas and different theorems which seem unrelated. In this paper we show the connections between these theories and present a new algorithm to recognize a comparability graph.

A comparability graph is an undirected graph $G = (V, E)$, $|V| = n$, $|E| = m$, in which every edge may be assigned a direction so that the resulting digraph is a partial order. To the best of our knowledge, literature so far knows mainly two solutions for this problem. The first solution is due to GOLUMBIC [7] and computes a *G-decomposition* recursively. This is a partition of the edges into so-called *implication classes* which define a transitive orientation. This algorithm runs in time $O(n \cdot m)$. The resulting orientation is transitive if the algorithm terminates successfully. The second solution with running time $O(n^2)$ was given by SPINRAD [11]. This algorithm computes a transitive orientation if there is such an orientation at all.

This is a fundamental difference to GOLUMBIC's algorithm. Whereas GOLUMBIC's algorithm finds a transitive orientation when it terminates successfully, SPINRAD's algorithm always computes an orientation of the graph. This orientation is transitive if there is such an orientation at all, however this has to be tested separately.

The difficult part of SPINRAD's method is the computation of the input data structure for the proper orientation process, the so-called *Modular Decomposition*, a recursively defined tree-representation of G. A restricted class of the Modular Decomposition, denoted as *cotrees*, arises in the context of *cographs*. In [2] an incremental algorithm is presented to construct a cotree in linear time.

Our algorithm is a natural extension of the cograph recognition algorithm in [2], it computes simultaneously a transitive orientation and a Modular Decomposition of a graph. The extension makes use of the condition which makes the cograph recognition algorithm fail. The main result is a structural improvement of the theory of comparability graphs. Our algorithm takes $O(n^2)$ time, which is a lower bound for algorithms relying upon a matrix representation of a graph. We show that modular decomposition and G-decomposition are not independent structures, they imply each other. Based on this perception, we develop new, shorter proofs of the algorithmically relevant theorems about comparability graphs and easily verifiable proofs for the correctness of the algorithms of SPINRAD and GOLUMBIC. Especially we show that SPINRAD's orientation is a special, efficient implementation of GOLUMBIC's algorithm.

1 Introduction

Recognition and orientation of a comparability graph is the solution of the *seriation problem* which has many applications in archaeology, psychology, political science and so on, see [9] and [8] for a survey. Furthermore, transitive orientation is a subproblem in several graph recognition problems, e.g. interval or permutation graphs, see [7]. Finally, for partial orders and therefore also for comparability graphs certain optimization problems, such as finding maximal cliques or coloring, can be efficiently solved, whereas they are NP-complete in the general case.

In order to start our analysis, we need some definitions. A *graph* $G = (V, E)$ is an ordered pair consisting of a set of vertices (nodes) $V = \{1, \ldots, n\}$ and a set of edges (arcs) E,

$$E \subseteq \otimes(V) \stackrel{\text{def}}{=} V \times V - \{(v, v) \mid v \in V\}.$$

An edge (u, v) is said to be *directed* from u to v, meaning that it starts in u and ends in v. For $A \subseteq E$ we define

$$\begin{aligned} A^{-1} &= \{(x, y) \mid (y, x) \in A\} \\ \hat{A} &= A \cup A^{-1} \\ A^2 &= \{(x, z) \mid \exists y \in V : (x, y) \in A \wedge (y, z) \in A\} \end{aligned}$$

The *adjacency set* $\operatorname{adj}(v)$ of a node $v \in V$ is given by

$$\operatorname{adj}(v) = \{w \in V \mid (v, w) \in E\}.$$

An *induced subgraph* G_B is a graph with vertex set $B \subseteq V$ and edge set

$$E_B = \{(a, b) \in E \mid a \in B \wedge b \in B\}.$$

It is said to be *induced* by B. Further $V(E')$ is the set of vertices which are *spanned* by E', $E' \subseteq E$, formally

$$V(E') = \{v \in V \mid \exists u \in V : (v, u) \in E' \vee (u, v) \in E'\}.$$

A graph $G = (V, E)$ is an *undirected* graph if and only if $E = \hat{E}$. An *orientation* of an undirected graph $G = (V, E)$ is a subgraph $T = (V, F)$ of G with

$$F \cap F^{-1} = \emptyset \text{ and } F \cup F^{-1} = E.$$

A *comparability graph* is an undirected graph $G = (V, E)$ which has a transitive orientation $T = (V, F)$, i.e.

$$F \cap F^{-1} = \emptyset, \quad F \cup F^{-1} = E, \quad F^2 \subseteq F.$$

Finally, a predicate P about G is *hereditary* if P holds true for every induced subgraph of G.

There are two different ways of computing a transitive orientation of an undirected graph. The first one was developed by GOLUMBIC in [6]. The underlying theory is based on the notion of "forcing" introduced by [4]. On the edges of a graph G we define a relation $\sim$ as follows:

$$(a,b) \sim (a',b') \quad \stackrel{\text{def}}{\Longleftrightarrow} \quad [a = a' \wedge (b,b') \notin E] \text{ or } [b = b' \wedge (a,a') \notin E]$$

We say that (a,b) *directly forces* (a',b'). As E does not contain self-loops, it is not difficult to verify that $(a,b) \sim (a,b)$ but also that $(a,b) \not\sim (b,a)$. The reflexive, transitive closure $\stackrel{*}{\sim}$ of $\sim$ is an equivalence relation. Its equivalence classes decompose the edges E of G into *implication classes*. Two edges (a,b) and (c,d) are in the same implication class if and only if there is a sequence of edges

$$(a,b) = (a_0,b_0) \sim (a_1,b_1) \sim \cdots \sim (a_k,b_k) = (c,d) \text{ with } k > 0.$$

We name this sequence a $\sim$-chain from (a,b) to (c,d) and we say that (a,b) *forces* (c,d) whenever $(a,b) \stackrel{*}{\sim} (c,d)$. Further we get

$$(a,b) \stackrel{*}{\sim} (a',b') \iff (b',a') \stackrel{*}{\sim} (b,a).$$

A $\sim$-chain *leads* from an edge $e = (a,b)$ to a node d if there is a $\sim$-chain

$$(a,b) = (a_0,b_0) \sim \cdots \sim (a_k,b_k) \sim (c,d), \quad k \geq 0.$$

$A_{(a,b)}$ is the implication class of G which contains the edge (a,b) and $(A_{(a,b)})_Y$ is the implication class of G_Y which contains the edge (a,b). Immediately by the definition of $\stackrel{*}{\sim}$ we get

Corollary 1. *Let A be an implication class of an undirected graph $G = (V,E)$ and furthermore let $T = (V,F)$ be a transitive orientation of G. Then we have*

$$A \cap F \neq \emptyset \quad \Rightarrow \quad A \subseteq F.$$

The central structure in GOLUMBIC's theory is the *G-decomposition*, a partition $E = \hat{B}_1 + \hat{B}_2 + \cdots + \hat{B}_k$ defined by

1. B_1 is an implication class of $G = G_1$.
2. B_i is an implication class of $G_i = G - (\hat{B}_1 + \hat{B}_2 + \cdots + \hat{B}_{i-1})$ for $i \geq 2$.

GOLUMBIC's algorithm follows directly from

Theorem 2. [6, 4] *For an undirected graph $G = (V,E)$ with G-decomposition $E = \hat{B}_1 + \hat{B}_2 + \cdots + \hat{B}_k$ the following statements are equivalent:*

1. *$G = (V,E)$ is a comparability graph.*
2. *$A \cap A^{-1} = \emptyset$ for all implication classes A of E.*
3. *$B_i \cap B_i^{-1} = \emptyset$ for $i = 1,\ldots,k$.*

Moreover, when these conditions hold, then $B_1 + B_2 + \cdots + B_k$ is a partial order[1].

[1] Note, in this paper a partial order always means a transitive acyclic digraph.

GOLUMBIC's algorithm finds an implication class B_i and removes it from G in an iterative fashion. In order to calculate B_i, we arbitrarily select an edge (x_i, y_i) and search all edges (x, y) with $(x_i, y_i) \overset{*}{\sim} (x, y)$ with a DFS-like procedure **Explore**. To find all edges forced by (x, y) we merge the lists adj(x) and adj(y) which are sorted in any given order. Since this work has to be done for all edges, the time bound of $O(n \cdot m)$ becomes obvious.

A different scheme is used by SPINRAD in [11]. He presents a scheme based on the Modular Decomposition which permits the orientation of a graph in a transitive way if such an orientation exists at all. The *Modular Decomposition* is a recursive partitioning of the vertices of a graph into maximal modules, where a subset M of V is a module if all vertices not in M are related either with all or with none of the vertices of M. The notion of a module has appeared under different names in literature, e.g. partitive set in [7], closed set, stable set, etc.

Definition 3. Let $G = (V, E)$ be an undirected graph. A set $M \subseteq V$ is called a *module* if the following property holds:

$$\forall z \in V - M: \quad \text{adj}(z) \cap M \neq \emptyset \;\Rightarrow\; \text{adj}(z) \cap M = M.$$

Furthermore, a module M is

- a *parallel* module if G_M is not connected,
- a *series* module if $\bar{G}_M$ is not connected, and
- a *neighborhood* module if both G_M and $\bar{G}_M$ are connected.

For a module M we call the set $V - M$ its *complement*. Finally, we will call a module M' of G_M with $M' \subset M$ a *submodule* of M.

The first two kinds of modules are trivial to orient, see [11]. Undirected graphs which contain only parallel or series modules are called *cographs*. There is an incremental algorithm to recognize them in linear time, see CORNEIL, PERL and STEWART [2]. The orientation of neighborhood modules is more difficult. In [11] two procedures are described which find a transitive orientation for a given neigborhood module M in time $O(|M|^2)$ if there is one. In this paper we show that this procedure can be extended to recognize the neighborhood module M simultaneously to its orientation. Next we will see that this recognition process for neighborhood modules can be integrated in the algorithm of [2], yielding a simple solution which computes both, a transitive orientation and a modular decomposition within running time $O(|V|^2)$. Our approach will combine the theories of GOLUMBIC and SPINRAD. First we show that Modular Decomposition and G-Decomposition imply each other. This leads to a simpler concept of the theory of comparability graphs, especially its algorithmically relevant parts.

This paper may be viewed as a trial to correct and simplify the theory of [10]. We do not improve the running time of previous solutions, but the conceptional simplicity. This paper improves the structure of known results, which was done by more papers during the history of science than finding completely new results. This must be true for transitive orientation especially, since there is no application of these algorithms which have a complete running time smaller than $O(n^2)$. Either we need to test for transitivity which is $\Omega(n^2)$ or we treat a graph and its complement which also implies $\Omega(n^2)$.

In the next section we will show how a graph can be decomposed in order to simplify the problem. Our algorithm will be given in the third section. The last section draws the conclusions from our work and shows where further work might be heading.

2 Implication classes and graph decomposition

We start with a technical fact which is a simpler version of GOLUMBIC's triangle lemma, see [7, Lemma 5.3].

Lemma 4. *Let A be an implication class of an undirected graph $G = (V, E)$ and let a, b, z be pairwise distinct vertices of G. For the edges (a,b), (a,z), (b,z) we find*

$$[(a,b) \in \hat{A} \wedge (a,z) \notin \hat{A} \wedge (b,z) \notin \hat{A}] \Rightarrow z \notin V(A).$$

Proof. Let (y,b) be an edge with $a \neq y \neq b$ and $(a,b) \sim (y,b)$. First we observe $y \neq z$ because otherwise there would be a contradiction to our assumption that $(b,z) \notin \hat{A}$. The same contradiction follows for $(z,y) \notin E$, hence

$$(z,y) \in E$$

must hold. From the definition of $\sim$ we infer

$$(a,b) \sim (y,b) \quad \Rightarrow \quad (a,y) \notin E$$

which implies

$$\underbrace{(a,z)}_{\notin \hat{A}} \sim (y,z) \quad \Rightarrow \quad (y,z) \notin \hat{A}.$$

Induction on the length k of an $\sim$-chain

$$(x_1, x_2) \sim (x_3, x_4) \sim \cdots \sim (x_{k-1}, x_k)$$

gives, using the above arguments repeatedly with x_{k-2} as a, x_{k-1} as b and x_k as y

(i) $\forall l, 1 \leq l \leq k: \quad (x_l, z) \in E \wedge (x_l, z) \notin \hat{A}$,
(ii) $z \notin \{x_1, \ldots, x_k\}$.

The lemma follows from part (ii).

The following lemmas point out the connection between modules and implication classes.

Lemma 5. [7, Proposition 5.10] *Let A be an implication class of G, then the set $V(A)$ is a module in G.*

Proof. Suppose this lemma is wrong. Then there is a vertex $v \in V - V(A)$ which is connected to some but not all vertices in $V(A)$. We define

$$R = \{ u \in V(A) \mid (u, v) \in E \}$$

the set of related and

$$U = \{ u \in V(A) \mid (u, v) \notin E \}$$

the set of unrelated vertices. Because $G_{V(A)}$ is connected (there must be a $\sim$-chain to any vertex), there exists an edge $(a, b) \in A$ with (without loss of generality) $a \in U$ and $b \in R$. Since $(a, b) \in \hat{A}$ we obtain the contradiction

$$[(a, b) \in \hat{A} \wedge (b, v) \in E \wedge (a, v) \notin E] \Rightarrow (b, v) \in \hat{A} \Rightarrow v \in V(A). \quad \blacksquare$$

Lemma 6. *Let M be a module of $G = (V, E)$ and furthermore let A be an implication class of G. Then*

$$\text{either } A \subseteq M \times M \text{ or } A \cap (M \times M) = \emptyset.$$

Proof. Definition 3 implies that for all edges $(x, y), (y, z) \in E$ with $x, y \in M$ and $z \in V - M$

$$[(x, y) \in E \wedge (y, z) \in E] \Rightarrow (x, z) \in E \Rightarrow (x, y) \not\sim (y, z)$$

from which the lemma follows.

The next lemma outlines the relationship between an implication class A and its spanning set $V(A)$.

Lemma 7. *Let A and B be implication classes of $G = (V, E)$. We obtain*

$$\hat{A} \cap \hat{B} = \emptyset \quad \Rightarrow \quad V(A) \neq V(B).$$

Proof. Suppose on the contrary that $V(A) = V(B)$. Now look at a vertex $c \in V(A)$. There must be an edge $(a, c) \in \hat{A}$ which is an edge of the $\sim$-chain leading to c. There is also an edge $(b, c) \in \hat{B}$ which is in the $\sim$-chain of B leading to c. Since $(a, c) \not\sim (b, c)$ we infer $(a, b) \in E$. By $\hat{A} \neq \hat{B}$ we get $(a, b) \notin \hat{A}$ or $(a, b) \notin \hat{B}$. Let without loss of generality $(a, b) \notin \hat{B}$, then Lemma 4 implies $a \notin V(B)$, in contradiction to $V(A) = V(B)$.

The next theorem will show the connection between a module M and the transitive orientation of the graph. First we give

Definition 8. Let $T = (V, F)$ be an orientation of an undirected graph $G = (V, E)$; further, let x be a vertex of a module M of G, so T is called *induced*[2] by the vertex x if the following condition holds:

$$\forall z \in V - M \; \forall v \in M : \quad (v, z) \in F \iff (x, z) \in F.$$

[2] This notion was first developed implicitly by GHOUILA-HOURI [5].

I.e. in the graph $T_{M+\{z\}}$ the vertex z is either a source or a sink, where source (sink) means a vertex with no (only) entering edges. Now we are ready to state the theorem which allows us to orient graphs.

Theorem 9. *Let M be a module of a graph $G = (V, E)$ and x a vertex in M. The following statements are equivalent:*

(1) *G is a comparability graph.*
(2) *G_M and $G_{(V-M)+\{x\}}$ are comparability graphs.*
(3) *G has a transitive orientation induced by x.*

Proof. First notice that "to be a comparability graph" is a hereditary property. This is clear for a partial order and therefore it follows for a comparability graph, too. Because of this, the implication (1) $\Rightarrow$ (2) is trivial by inheritance of transitivity. As the implication (3) $\Rightarrow$ (1) is trivial as well, we only need to show (2) $\Rightarrow$ (3).

Let $T' = (M, F')$ and $T'' = ((V - M) + \{x\}, F'')$ be transitive orientations of G_M and $G_{(V-M)+\{x\}}$. Furthermore, we define F''' as

$$F''' = \{ (y,z) \mid y \in M \wedge z \in V - M \wedge (x,z) \in F'' \} \\ \cup \{ (z,y) \mid y \in M \wedge z \in V - M \wedge (z,x) \in F'' \}.$$

Our claim is that $T = (V, F)$ with $F = F' \cup F'' \cup F'''$ is a transitive orientation induced by x. As T' and T'' are transitive orientations of G_M and $G_{(V-M)+\{x\}}$ respectively, it is clear that T is an induced orientation of G and we only need to show its transitivity. Consider three vertices $a, b, c \in V$ with $(a, b) \in F$ and $(b, c) \in F$. Four cases must be distinguished.

1. $(a, b) \notin F'''$ and $(b, c) \notin F'''$. Either $a, b, c \in M$ or $a, b, c \in V - M$ must hold. The edge (a, c) exists because F' and F'' are transitive.
2. $(a, b) \in F'''$ and $(b, c) \notin F'''$. Suppose $a \in M$, so $b, c \in V - M$ follows and we obtain
$$(a,b) \in F''' \;\Rightarrow\; (x,b) \in F''$$
by Definition 8, which implies
$$[(x,b) \in F'' \wedge (b,c) \in F''] \;\Rightarrow\; (x,c) \in F'' \;\Rightarrow\; (a,c) \in F''' \subseteq F,$$
by definition of F'' and F'''. In the other case, $a \in V - M$ and $b, c \in M$, the transitivity follows by
$$(a,b) \in F''' \;\Rightarrow\; (a,x) \in F'' \;\Rightarrow\; (a,c) \in F'''.$$
3. $(a, b) \notin F'''$ and $(b, c) \in F'''$. Analogous to (2).
4. $(a, b) \in F'''$ and $(b, c) \in F'''$. For $b \in M$ the definition of F''' implies
$$[(a,x) \in F'' \wedge (x,c) \in F''] \;\Rightarrow\; (a,c) \in F''$$
by the transitivity of F''. The reverse, $b \in V - M$, may not occur as we would find
$$(a,b) \in F''' \Rightarrow (x,b) \in F'' \text{ and } (b,c) \in F''' \Rightarrow (b,x) \in F'',$$
a contradiction to the definition of F'' as orientation of $G_{(V-M)+\{x\}}$. ∎

Note that, as the last case may never apply, the direction of any edge (u, v) with $u, v \in M$ has no influence outside M. Any transitive orientation of G_M combined with the constructed orientation of the other edges yields a transitive orientation of G. The construction of a transitive orientation for G_M is completely independent of the construction of an induced orientation for the rest of the graph.

The following Theorem 10 is identical with the induction step in the proof of Theorem 2. We give a new constructive proof for the necessity.

Theorem 10. *Let A be an implication class of a graph $G = (V, E)$. G is a comparability graph if and only if $G' = (V, E - \hat{A})$ and $G'' = (V, \hat{A})$ have a transitive orientation.*[3]

Proof. ($\Rightarrow$) By Lemma 5, the spanning set $V(A)$ of an implication class is a module. Therefore, there is a transitive orientation $T = (V, F)$ of G induced by $v \in V(A)$. Now suppose

$$T' = (V, F') \text{ with } F' = F - \hat{A}$$

is not a transitive orientation of G'. Then there exist vertices x, y, $z \in V$ with

$$(x, y) \in F' \wedge (y, z) \in F' \wedge (x, z) \notin F',$$

since T' is clearly an orientation of G'. Because T is transitive, (x, z) is an edge in E and $(x, z) \notin F'$ implies $(x, z) \in \hat{A}$. On the other hand, we get $(x, y) \notin \hat{A}$ and $(y, z) \notin \hat{A}$, since $(x, y), (y, z) \in F - \hat{A}$. From this we observe by Lemma 4

$$y \notin V(A)$$

and together with $(x, y), (y, z) \in F$ this is a contradiction to the definition of F as an induced orientation, because y would have to be either a source or a sink in $T_{M+\{y\}}$. Hence the assumption was wrong, therefore T' is a transitive orientation of G'.

Now consider $T'' = (V, A)$ where we suppose again that T'' is not transitive. Then there are nodes a, b, c with

$$(a, b) \in A \wedge (b, c) \in A \wedge (a, c) \notin A.$$

Now we have

$$\text{either } (a, c) \in A^{-1} \text{ or } (a, c) \notin \hat{A}.$$

In the first case, we consider, without loss of generality, a transitive orientation $T_1 = (V, F_1)$ of G with $(a, b) \in F_1$. From Corollary 1 we get

$$(a, b) \in (F_1 \cap A) \;\Rightarrow\; A \subseteq F_1 \;\Rightarrow\; [(b, c) \in F_1 \wedge (c, a) \in F_1].$$

On the other hand, by the transitivity of T_1, it follows

$$[(a, b) \in F_1 \wedge (b, c) \in F_1] \;\Rightarrow\; (a, c) \in F_1.$$

[3] The authors do not understand the proof of [7, Theorem 5.4, p.109] which treats the transitivity of G''.

Hence we obtain the contradiction

$$(a,c) \in F_1 \wedge (c,a) \in F_1.$$

Finally the case $(a,c) \notin \hat{A}$ must be considered. Let B be the implication class containing (a,c) in G. Then Lemma 4 implies

$$b \notin V(B).$$

With Theorem 9 we obtain a transitive orientation $T_2 = (V, F_2)$ of G induced by a. Let without loss of generality $(a,b) \in F_2$ then

$$\forall v \in V(B): \quad (v,b) \in F_2,$$

and in particular for $v = c$

$$(c,b) \in F_2.$$

But this is a contradiction, too, since Corollary 1 implies

$$(a,b) \in (F_2 \cap A) \;\Rightarrow\; A \subseteq F_2 \;\Rightarrow\; (b,c) \in F_2.$$

For this reason $T'' = (V, A)$ is transitive and this concludes the proof of the first part.

($\Leftarrow$) [7, p. 123] Let F' and F'' denote the transitive orientations of G' and G''. We need to show that the combination of F' and F'' yields a transitive orientation $T = (V, F)$ of G and we do this by considering two edges $(a,b), (b,c) \in F$. If both are in F' or F'' respectively, (a,c) exists by transitivity of F' or F''. Assume therefore that $(a,b) \in F'$ and $(b,c) \in F''$. Because (b,c) may not be in A, we get $(a,b) \not\Gamma (b,c)$ and therefore the edge (a,c) must be in E. Now suppose that $(a,c) \notin F$. Since F is an orientation, we observe $(c,a) \in F = F' \cup F''$. But for

$$[(c,a) \in F' \wedge (a,b) \in F'\,] \;\Rightarrow\; (c,b) \in F'$$

or

$$[(b,c) \in F'' \wedge (c,a) \in F''\,] \;\Rightarrow\; (b,a) \in F''$$

we get a contradiction to the definition of an implication class and $(a,c) \in F$ must hold. The argumentation is similar if $(a,b) \in F''$ and $(b,c) \in F'$. For this reason $F = F' \cup F''$ is indeed a transitive orientation of G.

3 The algorithm

With Theorem 10 our strategy becomes clear, namely: Find one implication class, orient it and its submodules recursively and then assign the induced transitive orientation to the entire graph. In order to find an implication class, we might proceed in the same way as GOLUMBIC in [6], but unfortunately this requires too much time. We thus perform a reduced version of GOLUMBIC's procedure **explore** to find the set $V(A)$ and use a different scheme to orient the edges.

We start with an arbitrary edge (u, v), orient it and proceed from there to assemble all the vertices which belong to the implication class $A_{(u,v)}$. These vertices must be related to some but not all other vertices in $V(A_{(u,v)})$. When we look at Algorithm 1 we see that lines 2–5 initialize this condition. The invariants of the loop 6–13 preserve this property; they are stated in the following lemma.

Lemma 11. *For the loop 6–13 the following invariants hold:*

$$\forall r \in R \; \forall b \in B: \quad (r, b) \in E \tag{1}$$

$$\forall u \in U \; \forall b \in B: \quad (u, b) \notin E \tag{2}$$

$$\forall b \in (B \cup Q) \; \exists b' \in B: \quad (x, y) \overset{*}{\sim} (b, b') \vee (x, y) \overset{*}{\sim} (b', b). \tag{3}$$

find_module1

```
(1)  procedure find_module1((x,y) ∈ E));
(2)      B ← {x,y};
(3)      U ← V − ({x} ∪ adj(x) ∪ {y} ∪ adj(y));
(4)      R ← adj(x) ∩ adj(y);
(5)      Q ← (adj(x) ∪ adj(y)) − (adj(x) ∩ adj(y));
(6)      while Q ≠ ∅ do
(7)          q ← a vertex in Q;
(8)          Q ← Q − {q};
(9)          B ← B + {q};
(10)         Q ← Q ∪ (R ∩ (V − adj(q)) ∪ (U ∩ adj(q)));
(11)         R ← R ∩ adj(q);
(12)         U ← U − adj(q);
(13)     od
(14)     return B;
(15) end find_module1
```

Algorithm 1

Proof. The invariants hold after the initialization in lines 2–5. We assume that they hold before the i-th execution of the loop in lines 6–13. Let B', Q', R' and U' denote the corresponding sets after this execution. Lines 7–8 do not change any of our invariants, line 9 adds the vertex q to B and so we have to ensure (1) and (2). As in line 11 only those vertices related with q stay in R, invariant (1) holds for R'. Similarly, in line 12 all vertices related with q are removed from U and (2) holds for U'. The vertices removed from R and U are added to Q in line 10. Let b be one of these vertices. By invariant (3) there is a vertex $q' \in B$ with

$$(x, y) \overset{*}{\sim} (q, q') \text{ or } (x, y) \overset{*}{\sim} (q', q).$$

One of the nodes q or q' fulfills the definition of b' for b, since we infer with invariant (1) for $b \in R$

$$(q, q') \sim (b, q')$$

and for $b \in U$ with invariant (2)

$$(q', q) \sim (b, q).$$

In both cases invariant (3) holds after the loop as well. So we have shown that all invariants hold at the end of the loop.

Corollary 12. *A call* **find_module1**$((x, y))$ *computes the set* $B = V(A_{(x,y)})$.

Proof. Consider the invariants proved in Lemma 11. In line 14 we observe from invariant (3) that

$$B \subseteq V(A_{(x,y)})$$

and from invariants (1) and (2) together with $Q = \emptyset$ it follows that B is a module at the end of the procedure, which with Lemma 5 implies

$$\forall x \in \underbrace{(U + R)}_{=V-B} : \quad x \notin V(A_{(x,y)}). \quad \blacksquare$$

After this procedure all the vertices spanned by the implication class of (x, y) have been found. Invariants (1) and (2) make clear that $V(A_{(x,y)})$ is a module and (3) shows that the edges between its vertices contain an implication class. Before looking at the implementation of **find_module**, some modification is necessary because we want to orient the edges of $A_{(x,y)}$ and not just find its spanning vertices.

Definition 13. Let A be an implication class of a graph G. The nonempty set $S \subset V(A)$ is called a *splitting set* of $V(A)$ if

$$\forall (x, y) \in E : \quad [x \in S \wedge y \in V(A) - S] \Rightarrow (x, y) \in \hat{A}.$$

Furthermore, we denote with

$$S(A) = (S \times (V(A) - S) \cap A)$$

the set of *splitting edges.*

Lemma 14. *Let S be a splitting set for the implication class A of a comparability graph $G = (V, E)$. Further, let A' be a subset of A containing the splitting edges of S. Then for given sets $V(A)$, S, and A' a call* **refine**$(V(A), S, A')$ *(see Algorithm 2) computes the implication class A and a partition $Part = \{Z_1, \ldots, Z_s\}$ of $V(A)$ with*

$$G_{V(A)} - \hat{A} = G_{Z_1} + \cdots + G_{Z_s}.$$

I.e. the sets Z_i are distinct submodules of $V(A)$ and all edges between Z_i and Z_j, $i \neq j$, belong to $\hat{A}$. The running time of the call **refine**$(V(A), S, A')$ *is bounded by*

$$O\Big(|\otimes(V(A)) - \bigcup_{1 \leq i \leq s} \otimes(Z_i)|\Big).$$

refine

```
(1)  procedure refine(V(A), S, A');
(2)      Part ← {S, V(A) − S}; (* Initial decomposition *)
(3)      out(S) ← V(A) − S; out(V(A) − S) ← S;
(4)      while ∃ Z ∈ Part with out(Z) ≠ ∅ do
(5)          choose a vertex v and remove it from out(Z);
(6)          Z' ← Z ∩ adj(v); Z'' ← Z − adj(v);
(7)          if ∅ ≠ Z' ≠ Z then
(8)              out(Z') ← out(Z) + Z'';
(9)              out(Z'') ← out(Z) + Z';
(10)             Part ← (Part − Z) + Z' + Z'';
(11)             forall (w, u) ∈ E with w ∈ Z' ∧ u ∈ Z'' do
(12)                 if (v, w) ∈ A'
(13)                 then A' ← A' ∪ {(u, w)};
(14)                 else A' ← A' ∪ {(w, u)};
(15)                 fi;
(16)             od;
(17)         fi;
(18)     od;
(19)     return (A', Part);
(20) end refine;
```

Algorithm 2

Proof. Let us consider a call **refine**$(V(A), S, A')$, where A' contains the splitting edges of S (see Algorithm 2). The procedure works by refining a partition *Part* of the vertices which is initialized with S and $V(A) - S$ where we have oriented all edges in between. For each element $Z \in$ *Part* we have a set $out(Z)$ which is initialized as follows:

$$out(S) = V(A) - S \text{ and } out(V(A) - S) = S.$$

This initialization fulfills the following invariants in the loop 4–18:

(i) For any set $Z_1 \in Part$ and any vertex $x \in (V(A) - (Z_1 + out(Z_1)))$ we have either $\text{adj}(x) \cap Z_1 = \emptyset$ or $\text{adj}(x) \cap Z_1 = Z_1$.

(ii) For any two sets $Z_1, Z_2 \in Part$ and for any edge $(x, y) \in E$ with $x \in Z_1, y \in Z_2$ the following holds: $Z_1 \neq Z_2 \Rightarrow (x, y) \in \hat{A}' \subseteq \hat{A}$.

The initialization before the first execution of the loop gives:

$$S + out(S) = (V(A) - S) + out(V(A) - S) = V(A)$$

which makes (i) trivial and (ii) is equivalent to the input conditions of the parameters $V(A), S$ and A'. Induction on the number of executions of the loop leads to the following cases:

1. $(\mathrm{adj}(v) \cap Z = \emptyset)$ or $(\mathrm{adj}(v) \cap Z = Z)$. In this case the set *Part* remains unchanged and (ii) holds because of the induction assumption. For (i) we only need to prove the case $Z_1 = Z$ and $x = v$ which again is equivalent to the assumptions.
2. $\emptyset \neq \mathrm{adj}(v) \cap Z \neq Z$. We replace Z in *Part* by $Z' = Z \cap \mathrm{adj}(v)$ and $Z'' = Z - \mathrm{adj}(v)$. If $Z_1 \notin \{ Z', Z'' \}$ and $x \in V(A) - (Z_1 + out(Z_1))$ then (i) does not change and it holds by the induction assumption. For $Z_1 \in \{ Z', Z'' \}$ and $x \in (V(A) - (Z + out(Z)))$, $x \neq v$, we find

$$\mathrm{adj}(x) \cap Z = \emptyset \quad \stackrel{Z=Z'\cup Z''}{\Rightarrow} \quad [\mathrm{adj}(x) \cap Z' = \emptyset] \wedge [\mathrm{adj}(x) \cap Z'' = \emptyset]$$

or

$$\mathrm{adj}(x) \cap Z = Z \quad \stackrel{Z=Z'\cup Z''}{\Rightarrow} \quad [\mathrm{adj}(x) \cap Z' = Z'] \wedge [\mathrm{adj}(x) \cap Z'' = Z''].$$

The induction assumption guarantees that either the first or the second implication holds, such that x fulfills (i) even after the execution of the loop 4–18. The last case is $Z_1 \in \{ Z', Z'' \}$ and $x = v$, but here (i) holds after the loop because of the definition of Z' and Z'' respectively. For invariant (ii) it is enough to show that any edge $(w, u) \in E$ with $w \in Z'$ and $u \in Z''$ must be in $\hat{A}$. The definition of Z' and Z'' in line 6 gives

$$[(v,w) \in E \wedge (w,u) \in E \wedge (v,u) \notin E] \;\Rightarrow\; (v,w) \sim (u,w).$$

With the induction assumption $(v, w) \in \hat{A}' \subseteq \hat{A}$ we find $(u, w) \in \hat{A}$ as claimed and we have proved both our invariants.

Next, we show that the loop 4–18 terminates because there can be no more than $|V(A)|$ elements in the partition and in each loop at least one v is removed from a set $out(Z)$. When the termination condition is reached $out(Z) = \emptyset$ holds for any set $Z \in Part$. Combining this with invariant (i) we find for all sets $Z \in Part$ and for all vertices $x \in (V(A) - (Z + out(Z)))$ that either

$$\mathrm{adj}(x) \cap Z = \emptyset \text{ or } \mathrm{adj}(x) \cap Z = Z.$$

By Definition 3, $out(Z) = \emptyset$ implies that any $Z \in Part$ is a module in $G_{V(A)}$. Together with Lemma 11 this shows for any three vertices $a \in V(A) - Z, b, c \in Z$

$$[(a,b) \in E \wedge (b,c) \in E] \;\Rightarrow\; (a,c) \in E \;\Rightarrow\; (a,b) \not\sim (c,b),$$

which implies

$$\hat{A} \cap (Z_i \times Z_i) = \emptyset.$$

Since this holds for any element of the final partition, we can conclude with invariant (ii) that $\hat{A} = A' \cup A'^{-1}$. By our precondition G is a comparability graph, therefore Theorem 2 part 2 requires $A \cap A^{-1} = \emptyset$ which implies $A' = A$ at the end of **refine**.

For the running time of the procedure **refine** we want to prove a result which is based on the resulting partition $Part = \{Z_1, \ldots, Z_s\}$. The implementation of

refine relies on an adjacency matrix representation of the graph G. The set E and the implication classes A_i may be stored in the same matrix, this allows the execution of the test "$(x, y) \in E$" or "$(x, y) \in A$" in constant time for any pair $x, y \in V$. With this representation we may state that one execution of the lines 11–16 costs time $O(|Z'| \cdot |Z''|)$ which is proportional to the maximal number of edges in $Z' \times Z''$. As the elements of the partition are never united, no subsequent execution of lines 11–16 will examine these edges again. If we sum over all executions of the lines 11–16, the running time is limited by the total number of examined edges. This is given by

$$| \otimes (V(A)) - \bigcup_{1 \leq i \leq s} \otimes (Z_i)|. \tag{4}$$

Note that none of the edges in $\otimes(Z_i)$ is ever examined because otherwise Z_i would not be an element of the final partition.

The matrix representation allows the implementation of line 6 as follows:

(6.1) $Z' \leftarrow \emptyset$; $Z'' \leftarrow \emptyset$;
(6.2) **forall** $w \in Z$ **do**
(6.3) **if** $(v, w) \in E$
(6.4) **then** $Z' \leftarrow Z' + \{w\}$
(6.5) **else** $Z'' \leftarrow Z'' + \{w\}$
(6.6) **fi**
(6.7) **od**

We now calculate the costs of line 6. As above, we count the number of times an edge is examined in line 6 and again we find that an edge (v, w) which was examined once in line 6 is not examined again, as v is removed from *out*(Z). For any set Z''' which contains w subsequently, $out(Z''') \subseteq out(Z) \cup Z \not\ni v$ holds. We find that in line 6 there will be no more than

$$O(| \otimes (V(A)) - \bigcup_{1 \leq i \leq s} \otimes (Z_i)|) \tag{5}$$

time spent.

So there only remain lines 8 and 9. We implement the sets *out*(Z) as linear lists. To find the lists *out*(Z') and *out*(Z'') we proceed as follows. We copy the list *out*(Z) once and reuse the list itself because we do not need it for Z anymore. These lists are extended, one by Z' and the other by Z''. The costs of copying are attributed to the costs of the newly created list. The total costs of the lines 8–9 (including line 2) is proportional to the total number of vertices created. But any vertex may serve exactly once as vertex v in line 6 and so this number is equal to the number of executions of the loop 4–18. As the costs of line 6 are at least $O(1)$ in each execution of the loop, they are larger than the costs of lines 8, 9 and 2 together which therefore are also less than the expression in (5). Now, summing up (4) and (5) leads to the claimed running time for **refine**.

We now state a condition for the existence of such a splitting set S which, when satisfied, allows its calculation.

Lemma 15. *Let A be an implication class of a graph G with $(x,y) \in A$. Then a splitting set S for the implication class exists if*

$$\exists z \in V(A) - \{x,y\} : \quad (x,z) \notin E \wedge (y,z) \notin E.$$

Proof. To find a splitting set with the procedure **find_module1** we base our method on the structure of that procedure. Consider a call of **find_module1** with (x,y) as the argument edge. B_i, U_i, R_i, Q_i and q_i indicates the contents of the corresponding variables after the i-th execution of line 7 in the loop 6–13. The index $i = \infty$ denotes the situation at the end of the procedure. By induction on i we define two mappings $d(v)$ and $dd(v)$ for all vertices $v \in V$. Initially for $i = 0$ let

$$d(v) = x \text{ and } dd(v) = y \quad \Longleftrightarrow \quad v \in \operatorname{adj}(x) - \operatorname{adj}(y) \tag{1}$$

and

$$d(v) = y \text{ and } dd(v) = x \quad \Longleftrightarrow \quad v \in \operatorname{adj}(y) - \operatorname{adj}(x). \tag{2}$$

For a vertex $v \in (Q_{i+1} - Q_i) \neq \emptyset,\ i > 1$, we define

$$d(v) = q_i \text{ and } dd(v) = d(q_i) \quad \Longleftrightarrow \quad v \in U_i \tag{3}$$

and

$$d(v) = d(q_i) \text{ and } dd(v) = q_i \quad \Longleftrightarrow \quad v \in R_i. \tag{4}$$

Case (3) is called a *forward implication* whereas situation (4) is called a *backward implication.* Directly from this definition we find for $x \neq v \neq y$

$$(v, d(v)) \sim (dd(v), d(v)), \tag{5}$$

and by induction on i

$$\forall v \in Q_i : \quad d(v), dd(v) \in B_i \tag{6}$$

and

$$\forall v \in B_i : \quad (v, d(v)) \in \hat{A}. \tag{7}$$

Let j be the largest index such that

$$\operatorname{adj}(q_j) \cap U_j \neq \emptyset. \tag{8}$$

This implies that in step j there was the last forward implication. For this index j we find $1 \leq j \leq |V(A)| - 3$, because of the preconditions of Lemma 15 there is a vertex $z \in V(A)$ which is neither related to x nor to y and thus must be in U_0. We next want to show that

$$S = U_j \cap \operatorname{adj}(q_j) \tag{9}$$

is a splitting set for V(A). To do this we have to show that all the edges between S and $V(A) - S$ lie in $\hat{A}$. First we show that

$$\forall w \in B_\infty - B_{j+1} : \quad d(w) \in B_{j+1}. \tag{10}$$

For $w \in S$ this holds by definition as q_j was just added to B_j the step before. For $w \in (Q_{j+1} \cap R_j)$ implies

$$d(w) = d(q_j) \overset{(6)}{\in} B_j \subseteq B_{j+1}.$$

Therefore (10) is true for $w \in Q_{j+1}$. Assume inductively that (10) is also correct for Q_i, $j + 1 \leq i$. Let $w \in Q_{i+1} - Q_i$. As all subsequent implications must be backward implications, the induction hypothesis implies

$$d(w) \overset{(4)}{=} d(q_i) \in B_{j+1}.$$

Hence (10) is correct for Q_{i+1} and this together with

$$Q_{j+1} \cup \cdots \cup Q_\infty = B_\infty - B_{j+1}$$

completes our argumentation.

Next we show that S indeed is a splitting set. Let us consider an arbitrary edge $(s, w) \in E$ with $s \in S$ and $w \in V(A) - S$. By definition of S we know

$$w \in B_\infty - B_j. \tag{11}$$

Now, one of the following three cases must be true:

1. $w = q_j$. Lemma 11 gives:
$$(s, q_j) \in \hat{A}. \tag{12}$$

2. $w \neq q_j \wedge d(w) \neq q_j$. From equation (10) we derive
$$d(w) \in B_j \overset{\text{Lemma 11}}{\Rightarrow} (s, d(w)) \notin E$$
which in turn implies
$$(w, s) \sim (w, d(w)) \quad \overset{(7)}{\Rightarrow} \quad (w, s) \in \hat{A}. \tag{13}$$

3. $w \neq q_j \wedge d(w) = q_j$. With (11) $w \in R_j$ must hold, (6) gives $dd(q_j) \in B_j$ and (8) implies $s \in U_j$. All this, together with Lemma 11 and (5) implies $(dd(q_j), s) \notin E$, $(dd(q_j), q_j) \notin E$ and $(dd(q_j), w) \in E$. So we find the $\sim$-chain
$$\hat{A} \overset{(7)}{\ni} (w, d(w)) = (w, q_j) \sim (w, dd(q_j)) \sim (w, s) \tag{14}$$
which shows that $(w, s) \in \hat{A}$ as claimed. ∎

Our next goal is the integration of **refine** in **find_module1** such that the combined running time does not exceed the time bound of **refine**. For reasons which will become clear later we have to avoid that $V(A)$ is a series or parallel module. In other words, in order to reach the desired running time, we must make sure that $V(A)$ is a neighborhood module. Fortunately there is a simple characterization of neighborhood modules by forbidden subgraphs.

Definition 16. A simple path $x_1, \ldots, x_s$, $s \geq 3$, is called *chordless* if

$$\forall 1 \leq i < j \leq s : \quad (x_i, x_j) \in E \iff i + 1 = j.$$

A P_4 is a chordless path with $s = 4$.

Theorem 17. *[1] For an undirected graph $G = (V, E)$, the following statements are equivalent.*

- *G is a cograph.*
- *G does not contain a neighborhood module.*
- *G does not contain a P_4.*

For this reason the required initialization is given by a P_4.

Lemma 18. *Let $x_1, \ldots, x_4$ be a P_4 in a comparability graph $G = (V, E)$. Further, let A be the implication class containing (x_1, x_2). Then a call to procedure* **find_module**$(V, x_1, \ldots, x_4)$ *computes $V(A)$, A and a partition of $V(A)$ in submodules $Part = \{Z_1, \ldots, Z_s\}$ of $V(A)$ with*

$$G_{V(A)} - \hat{A} = G_{Z_1} + \cdots + G_{Z_s}.$$

The running time of **find_module**$(V, x_1, \ldots, x_4)$ *is proportional to the sum of $|V|$ and the number of pairs in*

$$V(A) \times (V - V(A)) + \otimes(V(A)) - \bigcup_{1 \leq i \leq s} \otimes(Z_i).$$

Proof. The lines 2–32 represent a more detailed version of **find_module1**. For this reason the correct computation of $V(A)$ follows from Corollary 12. Next we see that $x_1 = x$, $x_2 = y$ and $x_4 = z$ fulfill the precondition of Lemma 15. For this reason a nonempty splitting set S is determined in line 33. Because lines 33–39. represent a direct implementation of the statements (12)–(14) of Lemma 15, the set A contains all splitting edges of S after line 39. But now in line 40 the sets $V(A)$, S and A fulfill the precondition of Lemma 14, which implies the correct computation of A and *Part*.

To analyse the running time of **find_module** we again use the matrix representation of the graph G. With this representation the loop in lines 7–10 looks like

forall $v \in V$ **do if** $(x, v) \in E \wedge (y, v) \notin E$ **then** ... **fi od;**

The lines 3–14 can be implemented similarly. We find that they can be executed in linear time, so no more than $O(|V|)$ time is spent in these lines. For the

find_module

```
(1)   procedure find_module(V, a chordless path x1, ..., x4);
(2)      x ← x1; y ← x2; S ← ø;
(3)      A ← {(x, y)}; B ← {x, y};
(4)      U ← V − ({x} ∪ adj(x) ∪ {y} ∪ adj(y));
(5)      R ← adj(x) ∩ adj(y);
(6)      Q ← (adj(x) ∪ adj(y)) − (adj(x) ∩ adj(y));
(7)      forall v ∈ adj(x) − adj(y) do
(8)         d(v) ← x; dd(v) ← y;
(9)         A ← A + {(x, v)};
(10)     od;
(11)     forall v ∈ adj(y) − adj(x) do
(12)        d(v) ← y; dd(v) ← x;
(13)        A ← A + {(v, y)};
(14)     od;
(15)     while Q ≠ ø do
(16)        if {x3, x4} ∩ Q ≠ ø
(17)        then choose x3 or x4 as q and remove it from Q
(18)        else choose any vertex q from the set Q and remove it
(19)        fi;
(20)        B ← B + {q};
(21)        if adj(q) ∩ U ≠ ø then qj ← q; S ← adj(q) ∩ U fi;
(22)        forall v ∈ ((R − adj(q)) ∪ (U ∩ adj(q))) do
(23)           if (q, v) ∈ E
(24)              then d(v) ← q; dd(v) ← d(q); U ← U − {v};
(25)              else d(v) ← d(q); dd(v) ← q; R ← R − {v};
(26)           fi;
(27)           Q ← Q + {v};
(28)           if (d(v), dd(v)) ∈ A then A ← A + {(d(v), v)};
(29)              else A ← A + {(v, d(v))};
(30)           fi;
(31)        od;
(32)     od;
(33)     forall s ∈ S do
(34)        forall w ∈ adj(s) ∩ (B − (S + {qj})) do
(35)           if (w, d(w)) ∈ A then A ← A + {(w, s)};
(36)              else A ← A + {(s, w)};
(37)           fi;
(38)        od;
(39)     od;
(40)     (A, Part) ← refine(B, S, A);
(41)     return (B, A, Part);
(42)  end find_module
```

Algorithm 3

loop 15–32 we use the same implementation. The costs of the i-th execution of this loop are proportional to the number of possible edges

$$(q_i, u), \quad u \in R_i \cup U_i.$$

The loop 22–31 is implemented as

forall $v \in R$ **do if** $(q,v) \notin E$ **then** ... **fi od;**
forall $v \in U$ **do if** $(q,v) \in E$ **then** ... **fi od;**

and we observe two facts. First of all, no edges from

$$(V - V(A)) \times (V - V(A))$$

are examined in lines 15–32 because

$$q_i \notin V - V(A)$$

holds for all q_i. A similar result is found for edges between vertices of submodules. Let $M \in Part$ be a submodule of $V(A)$, $M \subset V(A)$, then

$$|\{ x_1, \ldots, x_4 \} \cap M| \leq 1. \tag{15}$$

To verify (15), remember that $x_1, \ldots, x_4$ is a chordless path. This implies that all edges (x_i, x_{i+1}), $1 \leq i \leq 3$, are contained in $\hat{A} = \hat{A}_{(x_1,x_2)}$. From this we infer

$$\forall 1 \leq i \leq 3 : \quad \{ x_i, x_{i+1} \} \not\subseteq M,$$

because by Lemma 6 the contrary would imply $V(A) \subseteq M$ as opposed to our assumption $M \subset V(A)$. Without loss of generality, it remains to verify that the two other cases, namely

$$\{ x_1, \ldots, x_4 \} \cap M = \{ x_1, x_3 \} \text{ or } \{ x_1, \ldots, x_4 \} \cap M = \{ x_1, x_4 \}$$

do not apply. In the first case Definition 3 requires

$$\forall a, b \in V(A) - M : \quad \text{adj}(a) \cap M = \text{adj}(b) \cap M \tag{16}$$

which implies

$$\text{adj}(x_2) \cap \{ x_1, x_3 \} = \text{adj}(x_4) \cap \{ x_1, x_3 \},$$

a contradiction to the definition of $x_1, \ldots, x_4$ as a chordless path. The same contradiction arises in the second case and statement (15) is proved.

Therefore one of following three cases holds for M.

1. $\{ x_1, x_2 \} \cap M = \emptyset$. Then M is either contained in R_0 or in U_0 and from (16) we easily derive by induction on i the following invariant for loop 15–32

   $$\forall 1 \leq i \leq |V(A)| : \quad M \cap (B_i \cup Q_i) \neq \emptyset \;\Rightarrow\; M \subseteq (B_i \cup Q_i). \tag{17}$$

   This shows that during the current execution of the loop no edge in the graph G_M may be examined.

2. $\{x_1, x_2\} \cap M = \{x_1\}$. Then every node $v \in M$ is adjacent to x_2 but not adjacent to x_3. By (15) we have $x_2, x_3 \notin M$. Because of the choice $q_1 = x_3$ we get
$$(x_1, x_2) \sim (x_3, x_2) \sim (v, x_2) \Rightarrow M \subseteq Q_1.$$
3. $\{x_1, x_2\} \cap M = \{x_2\}$. Then every node $v \in M$ is adjacent to x_3 but not adjacent to x_4. With $q_1 = x_3$ and $q_2 = x_4$ we find
$$(x_1, x_2) \sim (x_3, x_2) \sim (x_3, x_4) \sim (x_3, v) \Rightarrow M \subseteq Q_2.$$

Therefore the costs of the loop 15–30 without the first two executions are bounded by
$$O(|V(A) \times (V - V(A))| + |\otimes(V(A)) - \bigcup_{1 \le h \le s} \otimes(Z_h)|). \tag{18}$$

The first two executions of the loop 15–32 are clearly bounded by $|V|$. Finally, in the lines 33–39 we do not consider edges in G_{Z_i} by definition of *Part*. Adding the costs of **find_module** without **refine** we obtain
$$O(|V|) + O(|V(A) \times (V - V(A))| + |\otimes(V(A)) - \bigcup_{1 \le h \le s} \otimes(Z_h)|).$$

This is also true for **refine**, see Lemma 14.

Next we present some properties of cographs. First we note that cographs have a unique tree representation, the so called *cotree*, see CORNEIL ET AL. [1]. The leaves of the cotree represent the vertices of the graph G. Internal nodes are labelled P or S such that the nodes on every path from the root to a leaf are labelled alternatively. This tree has the following interpretation. Two vertices $x, y \in V$ are adjacent if and only if the unique path from x to the root r of the tree meets the unique path from y to r at a node z with label S. Herewith, the node z is denoted as the *nearest common ancestor* of x and y. The cotree can be constructed in the following way. Since the vertex set V of a cograph $G = (V, E)$ is a (trivial) module, Theorem 17 requires that either G or $\bar{G}$ are not connected. In the former case we recursively build cotrees for the connected components of G and make them children of a newly created root r with label P. In the latter case replace G by $\bar{G}$ and label r with S.

In [2] an $O(n+m)$ algorithm is presented for the recognition of cographs and the construction of the corresponding cotrees. For us the most interesting aspect of this algorithm is, that it proceeds incrementally, i.e. the vertices are processed one at a time. Note that by inheritence, the induced subgraph $G_X, X \subset V$ of a cograph $G = (V, E)$ is also a cograph. The results of [2] can be summarized in

Theorem 19. *Let G_X, $X \subset V$, be a subgraph of an undirected graph $G = (V, E)$ and let v be vertex in $V - X$. Further, let G_X be a cograph with given cotree T_X. Then the cotree $T_{X \cup \{v\}}$ of $G_{X \cup \{v\}}$ can be determined in time $O(|\mathrm{adj}(v)|)$, if there exists one. Moreover, if $G_{X \cup \{v\}}$ is not a cograph, then a forbidden subgraph, namely a P_4, is found.*

Although the argumentation presented in [2] shows how to find P_4 only for one of several cases it is not difficult to extend the presented algorithm to find a P_4 in all cases.

Theorem 19 will be our central argument in developping our final recursion scheme in the procedure **orient** (see Algorithm 4). First we show that this procedure is correct, i.e. the output $T = (V, F)$ is a transitive orientation of the undirected graph $G = (V, E)$ if such an orientation exists at all. For this proof we use induction on the number n of vertices in G. Let us consider the main call of *orient*$(V, \emptyset, \emptyset)$. First we initialize F as an empty set. Next, we check for the case $V = \emptyset$, in order to avoid trivial calls. Furthermore, the parameter C and $\mathcal{T}_C$ are empty, too. The induction starts with $n = 1$. But this case is trivial, since there are no edges to orient. We assume for the induction that a call of **orient**$(V', C', \mathcal{T}_{C'})$ calculates a transtive orientation for all comparability graphs with $|V'| < n$, where C' is a subset of V' and $G_{C'}$ is a cograph with cotree $\mathcal{T}_{C'}$[4]. Then we have to show that **orient**$(V, \emptyset, \emptyset)$ finds a correct orientation of $G = (V, E)$.

In loop 3–31 we first test in an incremental way if G is a cograph. Thereby, an induced subgraph G_C, $C \subseteq V$, is calculated and this will be a cograph with cotree $\mathcal{T}_C$. If the condition in line 7 is always false then the loop 3–31 stops with $G_C = G$ and G is à cograph. In this case a transitive orientation of G is given by

Lemma 20. *[11] Let G_C be a cograph with a corresponding cotree $\mathcal{T}_C$. Since there are no edges between different children of the parallel nodes of a cotree, a transitive orientation $T = (C, F)$ of G_C is given by a transitive orientation of edges between children of series nodes. Let $c_1, \ldots, c_s$ be the children of a series node v, where the ordering on the children is arbitrary but fixed. Furthermore, let C_i the leaves in the subtree of c_i. Then we define*

$$\forall u \in C_i, v \in C_j, 1 \leq i, j \leq s : \quad (u, v) \in F \iff i < j.$$

The transitive orientation $T = (C, F)$ can be computed in linear time.

Otherwise, if the condition in line 7 is true then the subgraph $G_{C+\{v\}}$ is no longer a cograph. Now we may conclude with Theorem 17 that it contains a chordless path $x_1, \ldots, x_4$. By Lemma 18 the call **find_module**$(V, x_1, \ldots, x_4)$ determines the sets $A = A_{(x_1, x_2)}$, $V(A)$ and a partition *Part* of $V(A)$ in submodules. By Lemma 5 the set $V(A)$ is a module. This allows us to find an induced transitive orientation for G_V by Theorem 9 from orientations of $G_{V(A)}$ and $G_{(V - V(A)) + \{x\}}$. With Theorem 10 a correct orientation $T_{V(A)}$ of $G_{V(A)}$ is given by A, line 11, and a transitive orientation of $G_{V(A)} - \hat{A}$. Lemma 18 shows that $G_{V(A)} - \hat{A}$ consists of vertex disjoint components

$$G_{V(A)} - \hat{A} = G_{Z_1} + G_{Z_2} + \cdots + G_{Z_s}$$

[4] The parameters C and $\mathcal{T}_C$ are introduced to speed up the running time. As a matter of fact, they can be dropped during the discussion of correctness.

orient

```
(1)  procedure orient(V, C, T_C);
(2)  (* C ⊆ V, T_C is the cotree of G_C *)
(3)    while V − C ≠ ø do
(4)      choose a vertex v from V − C;
(5)      V ← V − { v };
(6)      compute the cotree T_{C+{v}};
(7)      if T_{C+{v}} undefined
(8)      then
(9)        let x_1, ..., x_4 be a chordless path in G_{C+{v}};
(10)       (V(A), A, Part) ← find_module(V, x_1, ..., x_4);
(11)       F ← F ∪ A;
(12)       forall Z ∈ Part with |Z| ≥ 2 do
(13)         C_Z ← Z ∩ C;
(14)         compute the cotree T_{C_Z};
(15)         orient(Z, C_Z, T_{C_Z});
(16)       od;
(17)       let x be one of the vertices in { x_1, ..., x_4 } ∩ C;
(18)       V' ← V − (V(A) − { x });
(19)       C' ← C − (V(A) − { x });
(20)       compute the cotree T_{C'};
(21)       if |V'| ≥ 2 then orient(V', C', T_{C'}) fi;
(22)       forall w ∈ V' − { x } do
(23)         if (x, w) ∈ F then
(24)           forall z ∈ V(A) − { x } do F ← F + { (z, w) } od fi;
(25)         if (w, x) ∈ F then
(26)           forall z ∈ V(A) − { x } do F ← F + { (w, z) } od fi;
(27)       od;
(28)       V ← ø; C ← ø;
(29)     else C ← C + { v }
(30)     fi;
(31)   od;
(32)   if |C| ≥ 2
(33)   then orient the cograph G_C with Lemma 20;
(34)     add this orientation to F
(35)   fi;
(36) end orient;
```

Algorithm 4

main

```
(1)  F ← ø;
(2)  if |V| > 1 then orient(V, ø, ø) fi;
(3)  return F;
```

Algorithm 5

with

$$Part = \{Z_1, Z_2, \ldots, Z_s\}.$$

Since $|Z_i| < |V|$ for $1 \le i \le s$, **orient**$(Z_i, \ldots)$ finds a transitive orientation of G_{Z_i} because of the induction assumption, line 12–16. As $\{x_1, \ldots, x_4\} \subseteq V(A)$, we obtain $|(V - V(A)) + \{x\}| < n - 1$. Hence our induction assumption is always true for $G_{(V-V(A))+\{x\}}$ and **orient**$((V - V(A)) + \{x\}, \ldots)$ always calculates a transitive orientation $T_{(V-V(A))+\{x\}}$. In lines 22–27 we find a transitive orientation for G_V which is induced by x and which contains both, $T_{V(A)}$ and $T_{(V-V(A))+\{x\}}$.

These arguments show that the recursion terminates, since we have $|V'| < |V|$ for any recursive call **orient**$(V', \ldots)$. Therefore, our scheme correctly calculates a transitive orientation of a graph G if such an orientation exists. It remains to determine the running time.

In the following, the set $Part = \{Z_1, \ldots, Z_s\}$ denotes the partition of vertices in $G_{V(A)} - \hat{A}$. Furthermore, let $z_i = |Z_i|$ and $k = |V(A)|$. A first cost category arises from the subgraph G_C. The costs of **find_module** can be split into three categories according to Lemma 18. Thereby, we ignore the recursive calls inside **orient**$(V, \ldots)$ for the moment. We distinguish the costs of initialization, for which we will give an upper bound, the costs of finding the set $V(A)$ and the costs to orient all edges of A.

$K_1(V)$ are the costs of building and orienting the cograph G_C, i.e. the costs of the line 6 if $G_{C+\{v\}}$ is a cograph and the cost of the line 33 if the condition "$|C| \ge 2$" is true.

$K_2(V)$ are the costs less than $O(n)$, [5]

$K_3(V)$ are the costs proportional to $|V - V(A)| \cdot |V(A)|$ and

$K_4(V)$ represents the costs of **find_module** which are proportional to neither $K_2(V)$ nor $K_3(V)$ in which case

$$K_4(V) = O(|\otimes(V(A)) - \bigcup_{1 \le i \le s} \otimes(Z_i)|).$$

For the call **orient**$(V, C, \mathcal{T}_C)$ the induced subgraph G_C of G always builds a cograph. Next note, during the execution of **orient**$(V, C, \mathcal{T}_C)$ we increase C by adding vertices or we split C into different parts for recursive calls, but we never remove elements from C. Therefore it holds: If $v \in C$ is an elemenet of V' for a recursive call **orient**$(V', C', \mathcal{T}_{C'})$ then v is also an element of C'. This shows that the line 6 is successfully executed at most once for a fixed vertex v over all calls of **orient**. By Theorem 19 one execution of line 6 is bounded by $O(n)$. Hence all successful executions of line 6 together take time $O(n^2)$. Because of Lemma 20 this time bound holds also for the lines 33–34 and we observe for the sum K_1 of all $K_1(V)$

$$K_1 = O(n^2). \tag{19}$$

[5] where n does not mean the number of vertices for this particular call, but the number of vertices in the original graph.

Now, we consider $K_2(V)$ which contains several contributions. The first one is clearly given by the costs of the main loop 3–31 without the lines 6–31. By Theorem 19 the next contribution to $K_2(V)$ arises from line 6 if $G_{C+\{v\}}$ is not a cograph. But this happens at most once for **orient**$(V,\ldots)$, because the condition in the following line 7 is true and with line 28 no further loop 3–31 will be executed. In addition, Theorem 19 implies if "$\mathcal{T}_{C+\{v\}}$ undefined" that in line 6 a chordless path $x_1,\ldots,x_4$ is calculated. From this we infer that the line 9 takes time $O(1)$. The call of **find_module** contributes time $O(|V|)$ to $K_2(V)$, see Lemma 18. Of course, the lines 11 and 13 are less expensive than $O(|V|)$. Finally, we come to the lines 14 and 20. In order to compute the cotree $\mathcal{T}_{C'}$ we

- duplicate the cotree $\mathcal{T}_C$,
- cut all branches leading into subtrees containing no leaves v, $v \in C$, and
- shrink all unary internal nodes.

This can be done in time proportional to the size of $\mathcal{T}_C$, i.e. in time $O(|C|) = O(n)$. It remains the line 14. Here we proceed in the same way as for $\mathcal{T}_C$, but we do not attribute the costs of calculation $\mathcal{T}_{C_Z}$ to the call **orient**$(V,\ldots)$ but to **orient**$(Z,\ldots)$. Since this addition to $K_2(Z)$ occurs at most once, the asymptotic costs of $K_2(Z) = O(n)$ does not change. Because of this the line 14 is free of charge in **orient**$(V,\ldots)$.

In order to sum up over $K_2(V)$, we show by induction on n that the number of (recursive) calls $or(n)$ of **orient** is bounded by

$$or(n) \leq n-1. \tag{20}$$

Since the case $n \leq 1$ is caught in the main routine, in the lines 12 and 21, respectively, no call of **orient** occurs as claimed. For $n \geq 2$ we assume that the claim is valid for all n', $1 \leq n' < n$. When we set $k = \sum_{i=1}^{s} \underbrace{|Z_i|}_{\stackrel{\text{def}}{=} z_i}$ we find for the number of calls

$$\begin{aligned} or(n) &\stackrel{\text{lines 15,21}}{=} or(n-k+1) + or(z_1) + \cdots + or(z_s) \\ &\stackrel{I.H.}{\leq} n-k+(z_1-1)+\cdots+(z_s-1) \\ &= n-k+k-s \\ &\leq n-1 \end{aligned}$$

which proves the proposition.

Now the sum K_2 of all costs $K_2(V')$ in the recursive calls **orient**(V') is easy to calculate. For any of these calls **orient**(V') we have $V' \subseteq V$ which leads to

$$K_2 = O(or(n) \cdot O(n)) \stackrel{(20)}{=} O((n-1) \cdot O(n)) \leq O(n^2) \tag{21}$$

The costs $K_3(V)$ of a call **orient**(V) are generated by the check of the edges between the module $V(A)$, and the vertices in $V - V(A)$. Since $|V| \geq 2$, we may state

$$K_3(V) = O(|(V - V(A)) \times V(A)|) = O(2 \cdot |(V - V(A)) \times (V(A) - \{v\})|) \tag{22}$$

for any vertex $v \in V(A)$. If costs of type $K_3(V)$ occur in a call **orient**(V), then V is split into

$$\textbf{orient}(V - V(A) + \{u\}, \ldots),\ u \in V(A)$$

and

$$\textbf{orient}(Z_1, \ldots), \ldots, \textbf{orient}(Z_s, \ldots).$$

The edges from $(V - V(A)) \times (V(A) - \{u\})$ are not examined by any call of **orient**$(V', \ldots)$, $V' \subset V$, as only edges from $V' \times V'$ are examined there. On the other hand, these edges cannot add to the costs of K_3 outside **orient**$(V', \ldots)$, as otherwise this call would not have occured. So we find that the sets $(V - V(A)) \times (V(A) - \{u\})$ are edge disjoint for all calls of **orient** which implies that the total costs for K_3 are bounded by

$$K_3 = O(|V \times V|) = O(n^2). \tag{23}$$

The costs $K_4(V)$ for a call of **orient**$(V, \ldots)$ are the remaining costs of **find_module** which are not part of $K_2(V)$ and $K_3(V)$. Lemma 18 shows that

$$K_4(V) = O(|P(V)|),$$

where the set of pairs $P(V)$ is given by

$$P(V) = \otimes(V(A)) - \bigcup_{1 \leq i \leq s} \otimes(Z_i). \tag{24}$$

The set $(A_{(x_1,x_2)})_V$ (in the following A_V) denotes the implication class containing the edge (x_1, x_2) which is chosen by **orient**$(V, \ldots)$. If we now consider the calls

$$\textbf{orient}(V', \ldots) \text{ and } \textbf{orient}(V'', \ldots)$$

with neither $V' \subseteq V''$ nor $V'' \subseteq V'$, we can say, based on our recursion scheme, that V' and V'' have at most one vertex in common. Therefore the sets $\otimes(V')$ and $\otimes(V'')$ have to be edge disjoint which immediately gives

$$P(V') \cap P(V'') = \emptyset.$$

On the other hand we have

$$\text{either } V' \subseteq V'' \text{ or } V'' \subseteq V'.$$

Without loss of generality, let $V' \subseteq V''$. Since $A_{V'} \cap A_{V''} = \emptyset$, it follows $V' \neq V''$ by Lemma 7. Then V' is a subset of some submodule Z_i of $V(A_{V''})$ and the set $\otimes(V(A_{V'}))$ is explicitly removed from $P(V'')$ in (24) and $P(V'), P(V'')$ are again edge disjoint. Thus all sets $P(V')$ are disjoint and the sum K_4 is bounded by

$$K_4 = O(|V \times V|) = O(n^2). \tag{25}$$

Theorem 21. *For an undirected graph $G = (V, E)$ the Algorithm 5 calculates an orientation $T = (V, F)$ of G, which is transitive if such an orientation exists at all. This goal is achieved in $O(n^2)$ time.*

Proof. The total costs of the call **orient**(V) are given by the sum

$$K_1 + K_2 + K_3 + K_4.$$

If we substitute the bounds found in (19),(21), (23) and (25) we find the claimed result.

$$K_1 + K_2 + K_3 + K_4 = O(n^2) + O(n^2) + O(n^2) + O(n^2) = O(n^2) \quad \blacksquare$$

4 Conclusions

We have presented an algorithm to reduce the comparability graph recognition problem to a check for transitivity in $O(n^2)$ time. This test may be done by calculating the transitive closure, but it is by no means obvious that this is the most efficient way to do this. So one of our current objectives is to find a more efficient test.

Our version of **orient** may be extended in such a manner that a Modular Decomposition, a G-Decomposition or both are generated. The G-Decomposition may be generated in $O(n^2)$ time regardless whether the graph is transitive or not, improving on GOLUMBIC's $O(n \cdot m)$ bound.

Moreover it is easy to see that GOLUMBIC's Theorem 5.29 in [7], in which it is stated that every G-decomposition has the same length, is a direct consequence of the fact that the Modular Decomposition of a graph is unique.

References

1. D. G. CORNEIL, H. LERCHS, AND L. STEWART BURLINGHAM, *Complement reducible graphs*, Discrete Applied Mathematics, 3 (1981), pp. 163–174.
2. D. G. CORNEIL, Y. PERL, AND L. K. STEWART, *A linear recognition algorithm for cographs*, SIAM J. Comput., 14 (1985).
3. A. COURNIER AND M. HABIB, *An efficient algorithm to recognize prime undirected graphs*, in Graph-Theoretic Concepts in Computer Science. 18th internationsl Workshop, E. W. Mayr, ed., Springer Verlag, Berlin, Germany, 1993, pp. 212–224.
4. T. GALLAI, *Transitiv orientierbare Graphen*, Acta Math. Acad. Sci. Hungar., 18 (1967), pp. 25–66.
5. A. GHOUILA-HOURI, *Caractérisation des graphes non orientés dont on peut orienter les arêts de manière à obtenir le graphe d'une relation d'ordre.*, C.R. Acad. Sci. Paris, 254 (1962), pp. 1370–1371.
6. M. C. GOLUMBIC, *The complexity of comparability graph recognition and coloring.*, Computing, 18 (1977), pp. 199–208.
7. ———, *Algorithmic Graph Theory and Perfect Graphs*, Academic Press, Inc., 1250 Sixth Avenue, San Diego California, 1980.
8. F. ROBERTS, ed., *Applications of combinatorics and graph theory to the biological and social sciences*, Springer-Verlag New York Berlin Heidelberg, 1989.
9. F. S. ROBERTS, *Graph theory and its applications to problems of society*, Society of industrial and applied mathematics, 1978.

10. J. P. Spinrad, *Two Dimensional Partial Orders*, PhD thesis, Princeton University, Oct. 1982.
11. ——, *On comparability and permutation graphs*, SIAM J. Comput., 14 (1985).

A Characterization of Graphs with Vertex Cover up to Five

Kevin Cattell[1] and Michael J. Dinneen[1,2]
kcattell@csr.uvic.ca mjd@lanl.gov

[1] Department of Computer Science, University of Victoria,
P.O. Box 3055, Victoria, B.C. Canada V8W 3P6
[2] Computer Research and Applications, Los Alamos National Laboratory,
M.S. B265, Los Alamos, New Mexico 87545 U.S.A.

Abstract. For the family of graphs with fixed-size vertex cover k, we present all of the forbidden minors (obstructions), for k up to five. We derive some results, including a practical finite-state recognition algorithm, needed to compute these obstructions.

1 Introduction

The proof of Wagner's conjecture by Robertson and Seymour (see [RS85, RS]), now known as the Graph Minor Theorem (GMT), has led to an explosion of interest in obstruction sets. Though the GMT is primarily of theoretical interest, our research group has been exploring applications of the theory. We have developed a system called VACS to help determine obstruction sets for certain graph families. One of these families, vertex cover, is the subject of this paper.

We make two main contributions in this paper. First, we present a linear-time algorithm that determines the vertex cover for the class of graphs with *bounded-pathwidth* (partial t-paths). There are related results as discussed below, but the algorithm we present has an important property of being *minimal* (defined in Section 3). It is this property which allows us to compute the second contribution of this paper: the obstruction sets for the first five fixed-parameter instances of the vertex cover problem.

The general problem of determining if a graph has a vertex cover of size k, with k part of the input, is well known to be $\mathcal{NP}$-complete [GJ79]. However, several $\mathcal{NP}$-complete problems have polynomial-time *fixed-parameter* versions for some fixed, problem-specific integer k. Vertex cover is an example of such a problem; the brute force approach of checking all k subsets of the vertices gives a crude $O(n^{k+2})$ algorithm. Alternatively, if a tree or path decomposition is available, determining the minimal vertex cover of a graph can be done linear time [ALS91]. In addition to our obstruction set characterizations, we present a practical, finite-state algorithm for path-decomposed graphs. Furthermore, by the facts (1) that path decompositions for fixed k can be found in linear time (see [Bod93, Klo93]) and (2) a pathwidth bound exists (see Theorem 11), we have an $O(n)$ algorithm. Interestingly, a direct fixed-parameter algorithm is presented in [DF] with the same complexity.

The rest of this paper is organized as follows. Section 2 formally defines the vertex cover problem, and introduces results and notation used in the paper. Section 3 presents our general vertex-cover algorithm and its minimal, finite-state variation for graphs of bounded pathwidth. Next, Section 4 contains results that reduce the amount of work needed to compute the obstructions sets. Finally, Section 5 presents the obtained obstructions sets.

2 Background

The most famous example of an obstruction set is found in Kuratowski's Theorem for planar graphs. It states that a graph G is planar if and only if G does not homeomorphically contain the complete bipartite graph $K_{3,3}$ or the complete graph K_5. This indicates the form of all obstruction set characterization of graph families; for some fixed graph family $\mathcal{F}$, $G \in \mathcal{F}$ if and only if G does not contain (under some partial order) any member of some set of graphs $\mathcal{O}(\mathcal{F}) = \{O_1, O_2, \ldots\}$.

For two graphs G and H, the graph H is a *minor* of the graph G if a graph isomorphic to H can be obtained from G by taking a subgraph and then contracting (possibly zero) edges. The GMT states that any set of finite graphs is a well-partial order under the minor order. A family $\mathcal{F}$ of graphs is a *lower ideal* (under the minor order) if $G \in \mathcal{F}$ implies that $H \in \mathcal{F}$ for any minor H of G. An *obstruction* O for a lower ideal $\mathcal{F}$ is a minor-order minimal graph not in $\mathcal{F}$. Hence, by using the GMT, a complete set of obstructions provides a *finite characterization* for any minor-order lower ideal.

For the remainder of this section, we formally define the vertex-cover lower ideals that are characterized in this paper along with other preliminary material. We first define the general vertex-cover decision problem (see [GJ79]) as follows:

Problem 1. Vertex Cover
Input: Graph $G = (V, E)$ and a positive integer $k \leq |V|$.
Question: Is there a subset $V' \subseteq V$ with $|V'| \leq k$ such that V' contains at least one vertex from every edge in E?

A set V' in the above problem is called a *vertex cover* for the graph G. The family of graphs that have a vertex cover of size at most k will be denoted by VC–k. For a given graph G, let $VC(G)$ denote the least k such that G has a vertex cover of cardinality k.

Lemma 2. The graph family VC–k is a lower ideal in the minor order.

Proof. Assume a graph $G(V, E) \in$ VC–k has a minimal vertex cover $V' \subseteq V$. If $H = G \setminus (u, v)$ for some $(u, v) \in E$ (edge deletion), then V' is also a vertex cover for H. Likewise, if $u \in V$ is an isolated vertex of G, V' also covers $H = G \setminus \{u\}$ (vertex deletion). For any edge $(u, v) \in E$, observe that $|\{u, v\} \cap V'| \geq 1$. Let w be the new vertex created from u and v in $H = G/(u, v)$ (edge contraction). Clearly, $V'' = (V' \cup \{w\}) \setminus \{u, v\}$ is a vertex cover of H with cardinality at most k. Since any minor of G can be created by repeating the above operations, VC–k is a lower ideal. □

Our computational system works with graphs of *bounded pathwidth*, which are defined below (see [CD] for detailed information). These graphs are somewhat related to the graphs of *bounded treewidth* characterized in [Ros73].

Definition 3. A *path-decomposition* of a graph $G = (V, E)$ is a sequence $X_1, X_2, \ldots, X_r$ of subsets of V that satisfy the following three conditions:

1. $\bigcup_{1 \le i \le r} X_i = V$,
2. for every edge $(u, v) \in E$, there exists an X_i, $1 \le i \le r$, such that $u \in X_i$ and $v \in X_i$, and
3. for $1 \le i < j < k \le r$, $X_i \cap X_k \subseteq X_j$.

The *pathwidth of a path-decomposition* $X_1, X_2, \ldots, X_r$ is $\max_{1 \le i \le r} |X_i| - 1$. The *pathwidth of a graph* G is the minimum pathwidth over all path-decompositions of G. Finding pathwidth is equivalent to many problems such as *gate matrix layout* and *vertex separation* [Möh90, EST87, KT92].

The family of graphs of pathwidth t or less, denoted by PW–t, can be represented by strings of operators from some *operator set*. There are many operators sets that can be used (e.g., for treewidth see [Wim87, ACPS91]), and we have chosen one of ours that eases both theory and implementation.

Our operator set Σ_t for bounded pathwidth graphs is defined by

$$\begin{aligned} \Sigma_t &= V_t \cup E_t \quad \text{where} \\ V_t &= \{ \textcircled{0}, \ldots, \textcircled{t} \} \quad \text{and} \\ E_t &= \{ \boxed{i\ j} \ : \ i, j \in V_t, i \neq j \}. \end{aligned}$$

The semantics of these operators on $(t+1)$-boundaried graphs are as follows:

- $\textcircled{i}$ Add an isolated vertex to the graph, and label it as the new boundary vertex i.
- $\boxed{i\ j}$ Add an edge between boundary vertices i and j (ignore if operation causes a self loop).

A graph described by a string of these operators is called a *t-parse*, and has an implicit labeled boundary ∂ of $t+1$ vertices. By convention, a t-parse always begins with the string $[\textcircled{0}, \textcircled{1}, \ldots, \textcircled{t}]$ which represent the edgeless graph of order $t+1$. When G is any t-parse and $Z \in \Sigma_t^*$ is any sequence of operators from the operator set Σ_t the *concatenation* of G and Z forms a new t-parsedenoted by $G \cdot Z$. The labeled boundary of the graph described by $G \cdot Z$ is different from G if the *extension* Z contains any 'new vertex' operators from V_t.

In [CD] it is shown that a graph G has a t-parse representation if and only if $G \in$ PW–t.

Example 4. A t-parse with $t = 2$ and the graph it represents. (The shaded vertices denote the final boundary.)

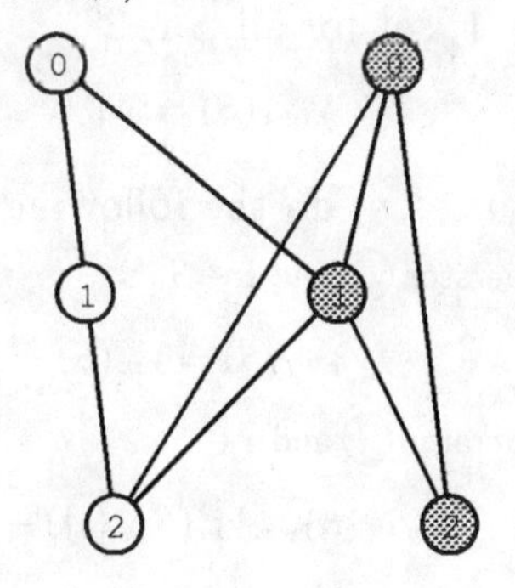

$$[⓪, ①, ②, \boxed{0\ 1}, \boxed{1\ 2}, ①, \boxed{0\ 1}, \boxed{1\ 2}, ⓪, \boxed{0\ 1}, \boxed{0\ 2}, ②, \boxed{0\ 2}, \boxed{1\ 2}]$$

3 Finite State Algorithm

In this section we give a practical, finite-state algorithm for the vertex cover problem on graphs of bounded pathwidth in t-parse form. This linear-time algorithm is a dynamic program that makes a single left to right scan of a t-parse $G_n = [g_1, g_2, \ldots, g_n]$. The computational process resembles a finite-state automaton in that it accepts words over the operator alphabet Σ_t. Let m be the current scan position of the algorithm on input G_n. The *state table* at operator g_m is indexed by each subset S of the boundary ∂. These 2^{t+1} different entries are defined as follows:

$$V_m(S) = \min\{\ |V'|\ :\ V' \text{ is a vertex cover of } G_m \text{ and } V' \supseteq S\}$$

Two important observations about the state table are:

1. For each boundary subset $S \in \partial$, $V_m(S)$ is a non-decreasing sequence of non-negative integers as m increases.
2. For any boundary subset $S \in \partial$ and any boundary vertex $i \notin S$, either $V_m(S) = V_m(S \cup \{i\})$ or $V_m(S) = V_m(S \cup \{i\}) - 1$.

The algorithm, given in Fig. 1, starts by setting the sizes for the minimal vertex covers on the empty graph $G_{t+1} = [⓪, ①, \ldots, \textcircled{t}]$, for all subsets S of the initial boundary ∂.

The type of the operator g_{m+1} (a vertex operator or an edge operator) determines how the state table is updated during the scan. The update of an entry for a specific subset of the boundary S is further broken up according to the relationship between S and the operator. These transitions are described in cases 1–4 of Fig. 1.

When the algorithm reaches the end of the t-parse, it has computed the minimum number of vertices needed for a vertex cover of G_n. This is because

I For $m = t+1$, set for all $S \in 2^\partial$

$$V_{t+1}(S) = |S|$$

II For $t+1 < m < n$, do the following cases:

Case 1: vertex operator ⓘ and $i \notin S$

$$V_{m+1}(S) = V_m(S)$$

Case 2: vertex operator ⓘ and $i \in S$

$$V_{m+1}(S) = V_m(S \setminus \{i\}) + 1$$

Case 3: edge operator $\boxed{i\ j}$, where $i \in S$ or $j \in S$

$$V_{m+1}(S) = V_m(S)$$

Case 4: edge operator $\boxed{i\ j}$, where $i \notin S$ and $j \notin S$

$$V_{m+1}(S) = \min\{V_m(S \cup \{i\}), V_m(S \cup \{j\})\}$$

III The vertex cover of G is

$$V_n(\emptyset)$$

Fig. 1. General vertex cover algorithm for t-parses.

the entry $V_n(\emptyset)$ contains the size of the smallest vertex cover that contains the subset $\emptyset$ of the boundary. As this is an empty condition, $V_n(\emptyset)$ is the size of the smallest vertex cover in G_n.

Theorem 5. For any t-parse $G_n = [g_1, g_2, \ldots, g_n]$, the algorithm in Fig. 1 correctly computes $VC(G_n)$.

Proof. If G is the empty graph then only steps I and III are executed and the correct result of $VC(G) = 0$ is returned. Assume that the algorithm is correct for all (prefix-) graphs of length m and less. We show that cases 1–4 of step II correctly update the state table, $V_{m+1}(S)$ for $S \in 2^\partial$.

Case 1: $g_{m+1} =$ ⓘ and $i \notin S$

Let V' be a witness vertex cover for $V_m(S)$. Since the new vertex created by g_{m+1} does not add any edges, V' is a vertex cover for G_{m+1}. Since $V' \supseteq S$ for G_m and $i \notin S$, $V' \supseteq S$ for G_{m+1}. Therefore, $V_{m+1}(S) \leq V_m(S)$.

Let V' be a witness vertex cover for $V_{m+1}(S)$. Since V' is minimal, the isolated vertex created by g_{m+1} is not in V'. Thus V' is a vertex cover for G_m. Since the property $V' \supseteq S$ is preserved, $V_m(S) \leq V_{m+1}(S)$.

Case 2: $g_{m+1} = ⓘ$ and $i \in S$

Let $S' = S \setminus \{i\}$ and V' be a witness vertex cover for $V_m(S')$. Now $W = V' \cup \{i\}$ is a vertex cover for G_{m+1} such that $W \supseteq S$. So, $V_{m+1}(S) \leq V_m(S') + 1$.

For the other direction, let V' be a witness vertex cover for $V_{m+1}(S)$. Since the new boundary vertex i does not help in any vertex cover of G_m, $V'' = V' \setminus \{i\}$ is a vertex cover for G_m such that $V'' \supseteq S$. Hence $V_{m+1}(S) \geq V_m(S') + 1$.

Case 3: $g_{m+1} = \boxed{i\ j}$ where $i \in S$ or $j \in S$

Let V' be a witness vertex cover for $V_m(S)$. Since the boundary is not changed by the edge operator g_{m+1} and $i \in S$ or $j \in S$, V' also covers the edges of G_{m+1}. Thus, $V_{m+1}(S) \leq V_m(S)$. If V'' is a vertex cover of G_{m+1} with $i \in S$ or $j \in S$, then V'' also covers the edges of G_m. So $V_{m+1}(S) \geq V_m(S)$.

Case 4: $g_{m+1} = \boxed{i\ j}$ where $i \notin S$ and $j \notin S$

Let $S' = S \cup \{i\}$ and V' be a witness vertex cover for $V_m(S')$. Since $V' \supseteq S$ is a vertex cover for G_{m+1}, as vertex i is in V', $V_{m+1}(S) \leq V_m(S')$. Likewise, if $S'' = S \cup \{j\}$, then $V_{m+1}(S) \leq V_m(S'')$. So $V_{m+1}(S) \leq \min\{V_m(S \cup \{i\}), V_m(S \cup \{j\})\}$.

Let V' be a witness vertex cover for $V_{m+1}(S)$. Since (i, j) is an edge, either $i \in V'$ or $j \in V'$. Thus $V' \supseteq S \cup \{i\}$ or $V' \supseteq S \cup \{j\}$. If $V' \supseteq S \cup \{i\}$, then V' is a vertex cover for G_m, and so $V_m(S \cup \{i\}) \leq V_{m+1}(S)$. Otherwise, $V' \supseteq S \cup \{j\}$, and $V_m(S \cup \{j\}) \leq V_{m+1}(S)$. Therefore, $\min\{V_m(S \cup \{i\}), V_m(S \cup \{j\})\} \leq V_{m+1}(S)$.

□

Example 6. The following table shows the application of the algorithm to the t-parse given in Example 4. As can been seen by examining the graph in Example 4, a minimum vertex cover has cardinality 3, which equals $V_{14}(\emptyset)$.

S \ m / g_m	3 –	4 $\boxed{0\ 1}$	5 $\boxed{1\ 2}$	6 ①	7 $\boxed{0\ 1}$	8 $\boxed{1\ 2}$	9 ⓪	10 $\boxed{0\ 1}$	11 $\boxed{0\ 2}$	12 ②	13 $\boxed{0\ 2}$	14 $\boxed{1\ 2}$
$\emptyset$	0	1	1	1	2	2	2	2	3	3	3	**3**
$\{0\}$	1	1	2	2	2	2	3	3	3	3	3	3
$\{1\}$	1	1	1	2	2	2	2	2	3	3	3	3
$\{2\}$	1	2	2	2	2	2	2	3	3	4	4	4
$\{0,1\}$	2	2	2	3	3	3	3	3	3	3	3	3
$\{0,2\}$	2	2	2	2	2	2	3	3	3	4	4	4
$\{1,2\}$	2	2	2	3	3	3	3	3	3	4	4	4
$\{0,1,2\}$	3	3	3	3	3	3	4	4	4	4	4	4

The following lemma shows that we can limit the vertex-cover membership algorithm to a finite number of possible configurations when testing for membership in VC–k.

Lemma 7. The algorithm in Fig. 1 is finite state for any fixed upper-bound k.

Proof. We show that for fixed k, there are only a finite number of possible states. Consider the state table entry for a boundary subset S. If $V_i(S)$ becomes $k+1$ for some i, then the monotonicity of $V_m(S)$ guarantees that $V_j(S) \geq k+1$ for all $j > i$. As we are only interested in knowing whether or not there exists a vertex cover of size k containing S, we can restrict $V_m(S)$ to be in $\{0, 1, 2, \ldots, k, k+1\}$. As there are 2^{t+1} entries in the state table, the number of states is bounded by $(k+2)^{2^{t+1}}$.

To make a fixed-parameter algorithm for VC–k, change any update function $V_{m+1}(S) = f(V_m)$ in the four cases with $V_{m+1}(S) = \min(f(V_m), k+1)$. It is straightforward to verify that this modified algorithm correctly computes the same state table except that any entry greater than $k+1$ is replaced by $k+1$. □

We define the *final state of a t-parse* G, denoted by V_G, to be the state of the finite-state algorithm when the algorithm terminates. For each boundary subset S, let $V_G(S)$ denote the S entry of V_G. If G and H are t-parses and $V_G = V_H$, it follows immediately that for any operator string $Z \in \Sigma_t^*$, we have $V_{G \cdot Z} = V_{H \cdot Z}$. This in turn implies that $G \cdot Z \in \mathcal{F} \Leftrightarrow H \cdot Z \in \mathcal{F}$ for all Z. That is, G and H *agree on all extensions.* The converse of this property is described by the following important definition.

Definition 8. A finite state algorithm for a family $\mathcal{F}$ is *minimal* if for any two t-parses G and H satisfying

$$G \cdot Z \in \mathcal{F} \Leftrightarrow H \cdot Z \in \mathcal{F} \text{ for all } Z \in \Sigma_t^*$$

then the final states of the algorithm are equal for the inputs G and H.

To show that our vertex-cover algorithm is minimal, we need to show that if G and H are t-parses, and G and H agree on all extensions, then $V_G = V_H$. We will show the contrapositive; that is, if $V_G \neq V_H$, then there exists an extension Z such that G and H do not agree on Z (that is, there exists Z such that either $G \cdot Z \in F$ and $H \cdot Z \notin F$, or $G \cdot Z \notin F$ and $H \cdot Z \in F$).

Before proving that our VC–k algorithm is minimal, we need the following lemma that provides us with an available boundary vertex for building such an extension Z.

Lemma 9. If G and H are t-parses such that $V_G(\partial) \neq V_H(\partial)$, then there exists an $S \subset \partial$ such that $V_G(S) \neq V_H(S)$.

Proof. Assume that $V_G(\partial) \neq V_H(\partial)$ is the only difference in the state table. The following three facts

1. $V_G(\partial \setminus \{i\}) = V_H(\partial \setminus \{i\})$ for all $i \in \partial$,
2. $V_G(\partial \setminus \{i\}) \leq V_G(\partial) \leq V_G(\partial \setminus \{i\}) + 1$ for all $i \in \partial$ and
3. $V_H(\partial \setminus \{i\}) \leq V_H(\partial) \leq V_H(\partial \setminus \{i\}) + 1$ for all $i \in \partial$

imply that

$$V_G(\partial) \le V_G(\partial \setminus \{i\}) + 1 = V_H(\partial \setminus \{i\}) + 1 \le V_H(\partial) + 1 \text{ for all } i \in \partial$$

and

$$V_G(\partial) \ge V_G(\partial \setminus \{i\}) = V_H(\partial \setminus \{i\}) \ge V_H(\partial) - 1 \text{ for all } i \in \partial \ .$$

After combining the above, $V_H(\partial) - 1 \le V_G(\partial) \le V_H(\partial) + 1$. So, without loss of generality, assume $V_G(\partial) = V_H(\partial) - 1 = d$. From this identity and facts 1 and 3 above (also see the partial state tables below), we must have $V_G(\partial) = V_G(\partial \setminus \{i\}) = d$ for all $i \in \partial$.

$$\begin{array}{c} \text{graph } G \\ \begin{array}{|c|c|} \hline V_G(\partial) & d \\ \hline V_G(\partial \setminus \{i\}) & \begin{array}{c} d-1 \\ \text{or} \\ d \end{array} \\ \hline \end{array} \end{array} \left.\begin{array}{c} < \\ \\ \end{array}\right\} = \left\{ \begin{array}{c} \text{graph } H \\ \begin{array}{|c|c|} \hline V_H(\partial) & d+1 \\ \hline V_H(\partial \setminus \{i\}) & \begin{array}{c} d \\ \text{or} \\ d+1 \end{array} \\ \hline \end{array} \end{array}\right.$$

This can happen if and only if each of the boundary vertices of G are attached to some non-boundary vertex. If not, then a vertex cover $V' \supseteq \partial$ of G would have a redundant vertex $i \in \partial$. The vertex cover created by eliminating vertex i from V' contradicts the value of $V_G(\partial \setminus \{i\})$. However, such a graph G can not exist since the last vertex operator can only have boundary vertex neighbors. Therefore, we can conclude that $V_G(\partial) = V_H(\partial)$ or there exists a $S \subset \partial$ such that $V_G(S) \ne V_H(S)$. □

Theorem 10. The finite-state algorithm in Fig. 1 is minimal for VC–k.

Proof. Let G and H be t-parses. As discussed above, we show that if $V_G \ne V_H$, then there exists an extension Z such that G and H do not agree on Z. Note that the theorem holds trivially if either one of G or H is not in $\mathcal{F}$ = VC–k by the empty extension $Z = [\,]$. If both $G \notin \mathcal{F}$ and $H \notin \mathcal{F}$ then $V_G = V_H = [k+1, k+1, \ldots, k+1]$ since $V_G(\emptyset) = k+1$ implies $V_G(S) = k+1$ for all $S \subseteq \partial$.

So suppose that $V_G \ne V_H$. Then without loss of generality, there is a boundary subset S with minimum cardinality such that $V_G(S) < V_H(S) < k+1$. Lemma 9 guarantees that $S \ne \partial$.

Let $\{v_1, v_2, \ldots, v_{|S|}\}$ be the boundary vertices in S. Pick a boundary vertex $i \notin S$ and any other boundary vertex $j \ne i$. Construct an extension Z as follows.

$$Z = [\textcircled{i}, \boxed{i\ v_1}, \textcircled{i}, \boxed{i\ v_2}, \ldots \textcircled{i}, \boxed{i\ v_{|S|}}, \overbrace{\textcircled{i}, \textcircled{j}, \boxed{i\ j}, \ldots, \textcircled{i}, \textcircled{j}, \boxed{i\ j}}^{k - V_H(S)+1 \text{ times}}]$$

The extension Z essentially forces the boundary vertices S to be covered while adding $k - V_H(S) + 1$ isolated edges. Now, VC$(H \cdot Z)$ is given by $V_H(S) + (k - V_H(S) + 1)$, which equals $k+1$, and so $H \cdot Z \notin \mathcal{F}$. However, VC$(G \cdot Z)$ is $V_G(S) + (k - V_H(S) + 1) < V_H(S) + (k - V_H(S) + 1) = k+1$, and so $G \cdot Z \in \mathcal{F}$. Therefore, the t-parses G and H do not agree on all extensions. □

4 VC–k Obstructions

Two ingredients suffice to compute obstruction sets for a lower ideal $\mathcal{F}$. First, we need to know a bound on the pathwidth (or treewidth) of the obstruction set. Such a bound always exists, as the obstruction set is finite. Given such a bound, we can compute all of the obstructions by restricting our search to a fixed pathwidth. Second, we require a minimal finite-state algorithm for $\mathcal{F}$ that operates on t-parses. An overview of how such an algorithm is used is given in Section 5. Our approach for computing obstruction sets is derived from the two theoretical approaches that appear in [FL89] and [LA91].

For vertex cover, we have both of these ingredients. A minimal finite-state algorithm was described in the previous section, and a pathwidth bound is shown later in this section.

In summary, computing the obstruction set for a vertex-cover family VC–k is as follows:

input: • pathwidth t
• minimal finite-state algorithm for VC–k, that operates on t-parses
output: • obstructions of pathwidth t

The next two subsections show what value of t is needed to get the complete set of obstructions $\mathcal{O}$(VC–k) for VC–k and why we can restrict our search to connected graphs.

4.1 Pathwidth of VC–k Obstructions

As discussed, a bound is required on the pathwidth of the obstruction set. That is, we need a result of the form *if* $G \in \mathcal{O}$(VC–k), *then* G *is of pathwidth* k' *or less.* For vertex cover, such a bound is easily obtained. We first show that the family VC–k is contained in the family PW–k. It follows from this that $\mathcal{O}$(VC–k) is contained in PW–$(k+1)$.

Theorem 11. The pathwidth of any member of VC–k is at most k.

Proof. For a given graph G of VC–k, let V' be a subset of of the vertices of size k that covers all edges. Denote the order of G by n. Let the vertices V of G be indexed by $1, 2, \ldots, n$ with the vertices $V \setminus V'$ coming first. We claim that $\{X_i \mid 1 \leq i \leq n-k\}$ where $X_i = V' \cup \{i\}$ is a path decomposition of G.

Since every vertex is either in V' or is in $V \setminus V'$ we have $\bigcup_{1 \leq i \leq n-k} X_i = V$. Let (u, v) be an edge of G. Since V' is a vertex cover, without loss of generality assume $u \in V'$. If also $v \in V'$ then any subset X_i contains both u and v. Otherwise, v must be indexed between 1 and $n-k$ and the subset X_v contains both u and v. Finally, note that for any $1 \leq i < j \leq n-k$ we have $X_i \cap X_j = V'$ (interpolation property satisfied). Thus, we have a path decomposition of pathwidth k. □

The above theorem can not be improved, as the complete graph K_{k+1} with pathwidth k is a member of VC–k.

Corollary 12. If $G \in \mathcal{O}(\text{VC}-k)$, then the pathwidth of G is at most $k+1$.

Proof. For any edge $(u,v) \in E(G)$, let $G' = G \setminus (u,v)$. Since G is an obstruction for VC–k, $G' \in$ VC–k and hence $\text{VC}(G') \leq k$ by Theorem 11. Let V' be a witness vertex cover for G'. Now $V = V' \cup \{u\}$ is a vertex cover for G of order at most $k+1$. Therefore, by Theorem 11 again, the pathwidth of G is at most $k+1$. □

4.2 Disconnected Obstructions

The number of obstructions we need to find can be reduced by some straightforward observations. The following are special cases of the more general results found in [CD].

Observation 13. Let C_1 and C_2 be graphs. Then $\text{VC}(C_1 \cup C_2) = \text{VC}(C_1) + \text{VC}(C_2)$.

Lemma 14. If $O = C_1 \cup C_2$ is an obstruction for VC–k, then C_1 and C_2 are obstructions for VC–k' and VC–k'', respectively, for some $0 < k', k'' < k$, with $k' + k'' = k$.

Hence we can restrict our attention to connected obstructions; any disconnected obstruction O of VC–k is a union of graphs from $\bigcup_{i=0}^{k-1} \mathcal{O}(\text{VC}-i)$ such that $\text{VC}(O) = k+1$.

Example 15. Since K_3 is an obstruction for VC–1, and K_4 is an obstruction for VC–2, the graph $K_3 \cup K_4$ is an obstruction for VC–3.

5 Results

The obstructions were computed using our VACS machinery (described in [CD]), which allows us to compute for any t all of the obstructions that have pathwidth at most t. However, tractability problems arise as t increases. As shown in the preceding section, we need to use pathwidth $k+1$ to obtain all of the obstructions for VC–k.

A brief description of the obstruction set computation is as follows. The set of all t-parses can be viewed as a tree, in which the parent of a length n t-parse G is the length $n-1$ prefix of G. The root of the tree is the empty graph $[⓪, ①, \ldots, \textcircled{t}]$. The minimality of the finite-state algorithm allows us to compute a 'pruning rule' for the tree. When this is done, the tree becomes finite, and the t-parses represented by the leaves form a set closely related to the VC–k obstruction set. This set, in fact, contains the obstruction set for VC–k.

A summary of our obstruction set computations (using a SPARC-2) for various VC–k families is shown in Table 1. The total graphs column shows the size of the pruned tree described above. In the minimal graphs column, the number of internal graphs plus obstructions is shown; that is, the leaves that are not

Table 1. Summary of obstruction set computation for vertex cover.

k	Elapsed time	Minimal graphs	Total graphs	Connected obstructions	Total obstructions
1	5 seconds	8	31	1	2
2	25 seconds	42	301	2	4
3	3 minutes	320	3,871	3	8
4	4 hours	4,460	82,804	8	18
5	6 days	121,228	3,195,445	31	56

obstructions have not been counted. The growth rate of the tree can be seen to be extremely high as k (and hence the pathwidth t) increases.

Besides the single obstruction K_2 for the trivial family VC–0, the connected obstructions for VC–k, $1 \leq k \leq 5$, are shown in Figs. 2–6. Some patterns become apparent in this set of obstructions. One such easily-proven observation is as follows.

Observation 16. For the family VC–k, both the complete graph K_{k+2} and the cycle C_{2k+1} are obstructions.

6 Conclusions

In this paper, we have presented the minor-order obstruction sets for the graph families VC–k, $1 \leq k \leq 5$. To calculate these obstructions, a minimal finite-state algorithm was developed. This algorithm was described and proven to be correct.

We are hopeful that this paper is the first in a succession of successful computations of obstruction sets for the plethora of graph families that are lower ideals in the minor order (e.g., k feedback vertex set and within-k-vertices of planarity). The system used to compute the obstruction sets is constantly improving, allowing us to reach higher pathwidths and more complicated problems.

Acknowledgement

The authors wish to thank Mike Fellows for introducing us to this exciting field of study and Todd Wareham for comments on an earlier version of this paper.

References

[ACPS91] Stefan Arnborg, Derek G. Corneil, Andrzej Proskurowski, and Detlef Seese. An algebraic theory of graph reduction. In *Proceedings of the Fourth Workshop on Graph Grammars and Their Applications to Computer Science*, volume 532, pp. 70–83. Lecture Notes in Computer Science, Springer-Verlag, 1991. To appear in Journal of the ACM.

[ALS91] Stefan Arnborg, Jens Lagergren, and Detlef Seese. Easy problems for tree-decomposable graphs. *Journal of Algorithms*, **12** (1991), pp. 308–340.

[Bod93] Hans L. Bodlaender. A linear time algorithm for finding tree-decompostions of small treewidth. In *Proc. 25th Annual ACM Symposium on Theory of Computing*. ACM Press, 1993.

[CD] Kevin Cattell and Michael J. Dinneen. VLSI Automated Compilation System – obstruction set computations (technical notes). In preparation.

[DF] Rod Downey and Michael R. Fellows. Parameterized computational feasibility. In P. Clote and J. Remmel, editors, *Feasible Mathematics II*. Birkhauser. To appear.

[EST87] J. Ellis, I. H. Sudborough, and J. Turner. Graph separation and search number. Report DCS-66-IR, Dept. of Computer Science, University of Victoria, August 1987. To appear in *Information and Computation*.

[FL89] Michael R. Fellows and Michael A. Langston. An analogue of the Myhill-Nerode Theorem and its use in computing finite-basis characterizations. In *Proc. Symposium on Foundations of Computer Science (FOCS)*, pp. 520–525, 1989.

[GJ79] Michael R. Garey and David S. Johnson. *Computers and Intractability: A Guide to the Theory of NP-Completeness*. W. H. Freeman and Company, 1979.

[Klo93] Ton Kloks. *Treewidth*. Ph.D. dissertation, Dept. of Computer Science, Utrecht University, Utrecht, the Netherlands, 1993.

[KT92] András Kornai and Zsolt Tuza. Narrowness, pathwidth, and their application in natural language processing. *Discrete Applied Mathematics*, **36** (1992), pp. 87–92.

[LA91] Jens Lagergren and Stefan Arnberg. Finding minimal forbidden minors using a finite congruence. In *Proceedings of the 18th International Colloquium on Automata, Languages and Programming*, volume 510, pp. 533–543. Springer-Verlag, Lecture Notes in Computer Science, 1991.

[Möh90] Rolf H. Möhring. Graph problems releted to gate matrix layout and PLA folding. In G. Tinhofer, E. Mayr, H. Noltemeier, and M. Syslo, editors, *Computational Graph Theory*, pp. 17–51. Springer-Verlag, 1990.

[Ros73] Donald J. Rose. On simple characterizations of k-trees. *Discrete Mathematics*, **7** (1973), pp. 17–322.

[RS] Neil Robertson and Paul D. Seymour. Graph Minors. XVI. Wagner's conjecture. To appear *Journal of Combinatorial Theory, Series B*.

[RS85] Neil Robertson and Paul D. Seymour. Graph Minors – A Survey, In *Surveys in Combinatorics*, volume 103, pp. 153–171, Cambridge University Press, 1985.

[Wim87] T. V. Wimer. *Linear algorithms on k-terminal graphs*. Ph.D. dissertation, Dept. of Computer Science, Clemson University, August 1987.

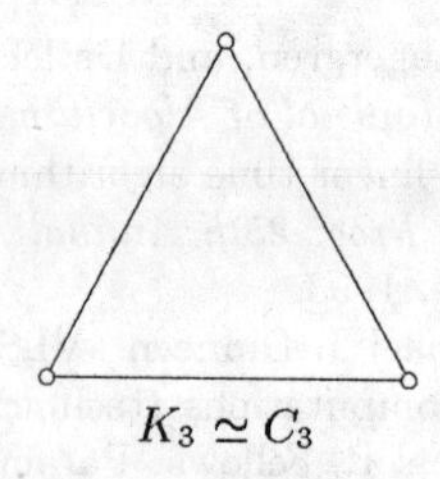

Fig. 2. Connected obstructions for Vertex Cover 1.

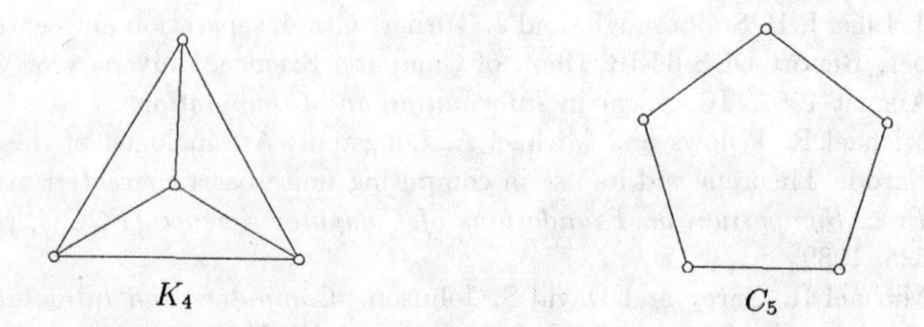

Fig. 3. Connected obstructions for Vertex Cover 2.

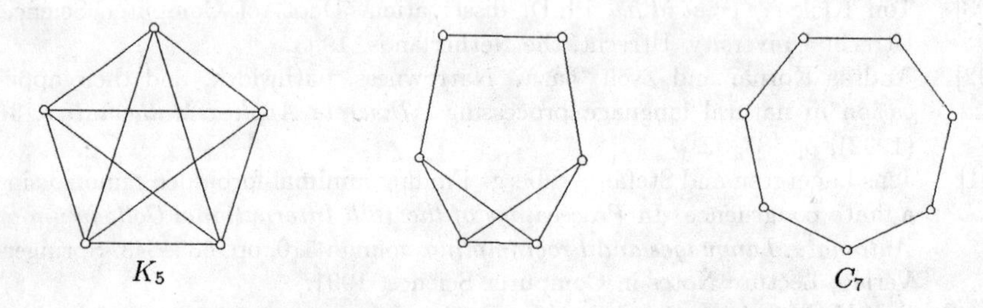

Fig. 4. Connected obstructions for Vertex Cover 3.

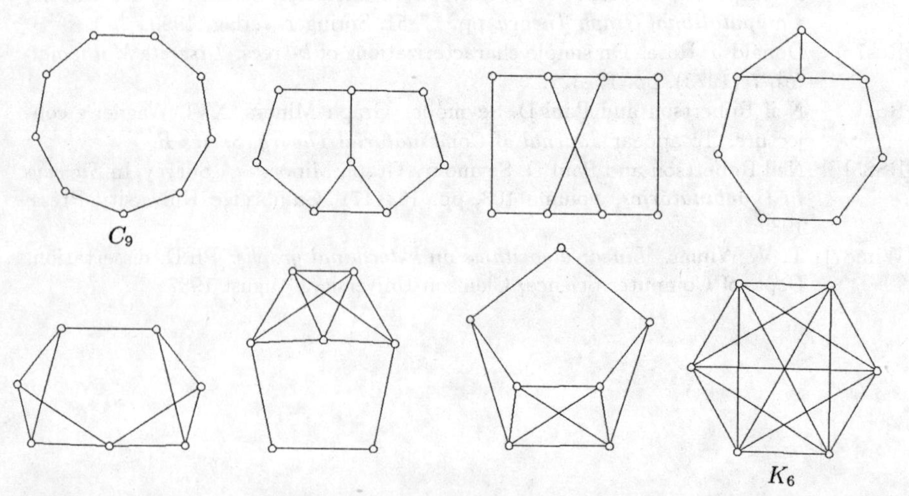

Fig. 5. Connected obstructions for Vertex Cover 4.

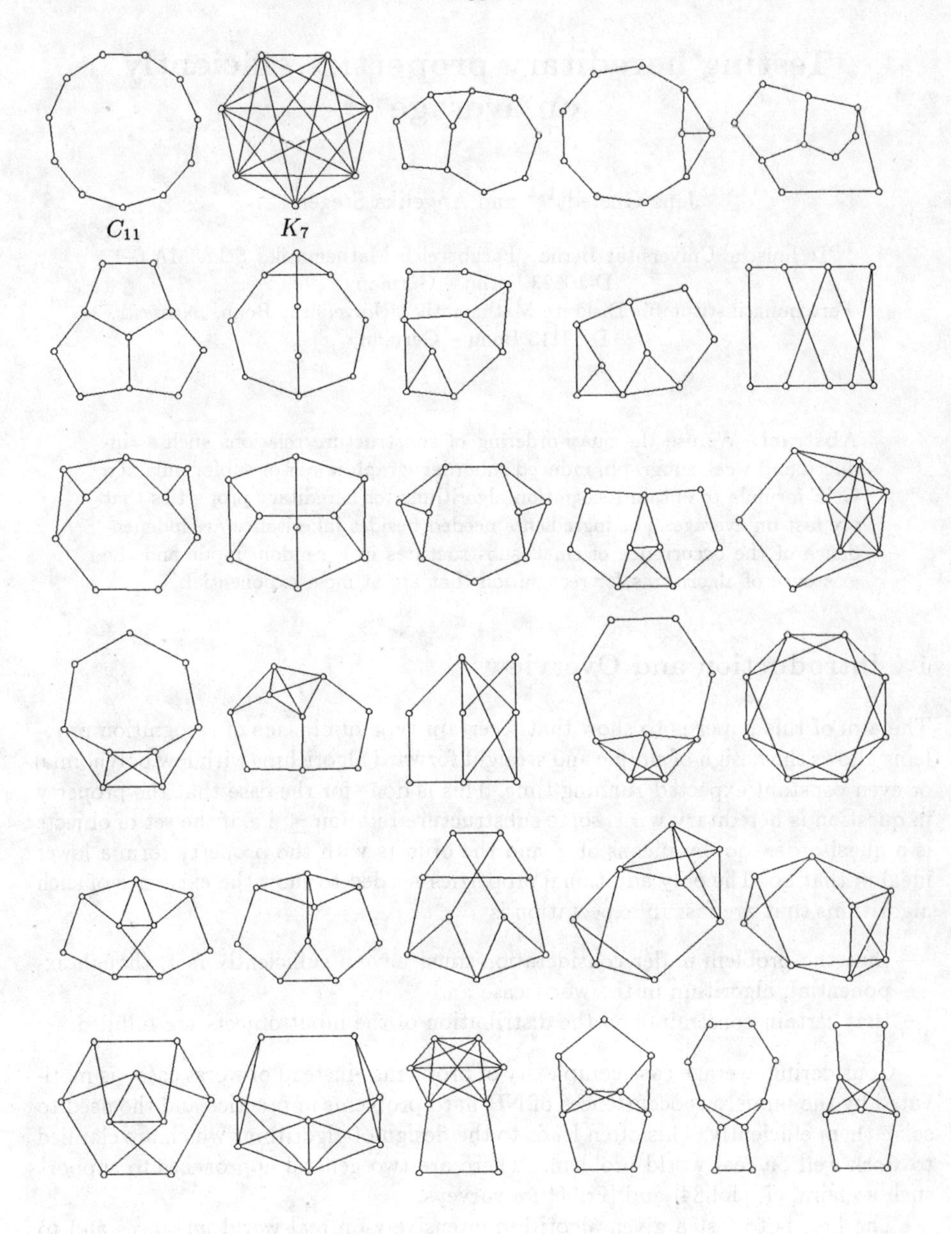

Fig. 6. Connected obstructions for Vertex Cover 5.

Testing hereditary properties efficiently on average*

Jens Gustedt[1]** and Angelika Steger[2]***

[1] Technische Universität Berlin, Fachbereich Mathematik, Sekr. MA 6–1, D-10623 Berlin – Germany
[2] Forschungsinstitut für Diskrete Mathematik, Universität Bonn, Nassestr. 2, D-53113 Bonn – Germany

Abstract. We use the quasi-ordering of substructure relations such as induced and weak subgraph, induced suborder, graph minor or subformula of a CNF formula to obtain recognition algorithms for hereditary properties that are fast on average. The ingredients needed besides inheritance are independence of the occurrence of small substructures in a random input and the existence of algorithms for recognition that are at most exponential.

1 Introduction and Overview

The aim of this paper is to show that a certain type of classes of recognition problems allows the design of simple and straight forward algorithms with low polynomial or even constant expected running time. This is done for the case that the property in question is hereditary w.r.t. some substructure relation $\preceq$, i.e. if the set of objects is a quasi-order, qo, by means of $\preceq$ and the objects with the property form a lower ideal in that qo. The only additional properties needed to show the existence of such algorithms that are fast in expectation is

- that the problem under consideration must have a sufficiently fast, though exponential, algorithm in the worst case and
- that certain constraints on the distribution of the input objects are fulfilled.

Considering average case complexity of problems –instead of worst case– is motivated by the numerous occurrences of **NP**-hard problems in practice and the need to solve them efficiently. This often leads to the design of algorithms which are claimed to work well on real world problems. There are two general approaches to support such a claim, cf. [Joh84] and [Fri90] for surveys.

The first is to test a given algorithm intensively on real world instances and to view the results as an evidence of its good performance. Of course no guarantee can be given for such a procedure.

The second is to prove some theorem stating that the algorithm performs good on average, cf. [BFF85, BK79, DF89, FP83, GPB82, HTL91, Iwa89, PS92a, Wil84].

---

* Part of this work has previously been published in the first author's Ph.D. thesis, [Gus92].

** Supported by the "Graduiertenkolleg Algorithmische Diskrete Mathematik", DFG grant We 1265/2-1.

*** Supported by DFG grant Pr 296/2-1.

To formulate such a statement precisely it is necessary to specify the probability distribution over which the average is taken. This, however, is usually not possible for distributions occurring in practice. In addition the difficulty in the analysis of complicated distributions restricts the possible choices even further.

So as a very common approach in average case analysis only "natural" distributions are considered. One example is the uniform distribution on graphs, which can be realized by considering a random graph in which every edge appears with a certain probability independently of all other edges.

We show that such distributions allow the design of simple and straightforward algorithms for a large variety of problems. The purpose of the algorithms obtained may be seen under two different aspects.

First, the average case complexity of any given algorithm and distribution must be judged in comparison to the algorithms obtained by this straightforward approach. It should only be claimed that an algorithm behaves well on average, if its performance is at least as good as that of the algorithm obtained by our approach.

Second, our results provide simple tools to improve known algorithms such that they behave well on average. This means in particular, that when designing an algorithm it is not necessary to consider its average case behavior — if a good average complexity is needed, it can easily be achieved in a last "fine-tuning" step of the implementation.

In the last years a similar approach in structural graph theory has proven to be very fruitful in showing existence of efficient worst case algorithms for a large class of problems, cf. e.g. [RS85b, FL85, FL88, Fel89, Möh90, RS85a, FL92]. The Graph Minors Theory developed by Robertson and Seymour gives such algorithms for graph properties that are hereditary with respect to the graph minor relation. It is based on the fact that the properties under consideration are characterized by a finite number of minimal obstructions.

The fact that this set of obstructions is finite is crucial for the design of an algorithm testing such a property. It reduces to show that each of the obstructions can be tested efficiently. When attempting to generalize such an approach to other combinatorial structures one immediately realizes that most of the structures of interest are not equally well behaved – the sets of minimal obstructions are usually not finite.

The present paper is designated to keep the concept of designing fast algorithms for hereditary properties of such structures as well but it restricts the attention to algorithms that are fast on average. With "fast on average" we mean here that they are fast on a random choice of a designated input space.

The key observation for such algorithms is the following. If the probability p is low that a fixed obstruction H of property $\mathcal{E}$ is not contained in the input then most instances may not have property $\mathcal{E}$. So testing for the obstruction H in a first step already gives the correct answer for most of the possible instances. Then any exact algorithm $T_{\mathcal{E}}$ can be used to investigate those instances that do not contain the obstruction.

If we assume

- that $\mathrm{p}(n)$ is a decreasing function, $o(1/f(n))$, say, where $f(n)$ tends to infinity and n denotes the input size of an instance and

– that $T_{\mathcal{E}}$ has a running time bounded by $O(f(n))$

$T_{\mathcal{E}}$ in total only contributes an additive constant to the expected running time. Clearly, the time needed by an algorithm composed of two parts in such a way is dominated by the test for the obstruction.

In this paper we will develop the three parts of such an approach, namely

(i) to show that a fixed obstruction H of a property $\mathcal{E}$ occurs with high probability
(ii) to develop an algorithm that is a test for a given obstruction H and that is fast on average
(iii) to design an exact algorithm for $\mathcal{E}$ whose running time is sufficiently small compared to the probability that the obstruction H does not occur.

We will do that in a general setting if possible, but we will also consider some special examples of combinatorial structures. Among these will be several classes of graphs equipped with three different relations, namely the induced and weak subgraph relation, and the graph minor relation, and classes of orders together with the induced suborder relation.

The paper is organized as follows. Section 2 provides the necessary terminology on order relations and hereditary properties and then gives some basic observations about average time complexity. In Section 3 we illustrate our general principle of constructing fast expected time algorithms on hereditary graph properties. The results of this section provide straightforward $O(1)$ expected time algorithms for several **NP**-complete problems. One of them is deciding whether a graph is k-colorable, a problem for which Wilf previously gave such an algorithm, [Wil84]. We also obtain a test for perfectness which has constant expected running time.

In Section 4 we develop a general approach for sampling small substructures. We give a simple algorithm to do so if the substructures can be obtained from sets of subsets of bounded cardinality k of a given structure V, where k is a fixed constant. As a first example for this technique we give efficient algorithms for properties in some classes of orders.

Then we illustrate our method at an example not coming from graph theory. In Section 5 we give constant time algorithms for testing CNF formulas, a problem which previously had been extensively treated within the theory of average case analysis (cf. e.g. [GPB82], [FP83] and [Iwa89]). In Section 6 we investigate classes of sparse graphs and the three relations on graphs mentioned above. For the weak and for the induced subgraph relation we achieve a tradeoff between the worst case complexity of an algorithm testing a given property and the edge probability function for which we can construct efficient average case algorithms. For the Graph Minor relation we design algorithms with linear expected running time for practically every hereditary property and every edge probability function.

The density of the graphs will vary from edge probability $1/2$ to graphs with only a constant number of edges per vertex. The properties that have fast algorithms will get more and more restricted.

2 Terminology and Basic Results

We want to give a general method to speed up the average time complexity of algorithms. Here "speeding up" means the following. Assume we have an algorithm $T_{\mathcal{E}}$ to test a certain property $\mathcal{E}$. This algorithm might be expensive. Our aim is to avoid a call to $T_{\mathcal{E}}$ by putting a cheaper algorithm $S_{\mathcal{E}}$ in front.

$S_{\mathcal{E}}$ should give one of three possible answers, **true** — the input does have property $\mathcal{E}$, **false** — the input does not have property $\mathcal{E}$, or **maybe** — the input might or might not have property $\mathcal{E}$. We will refer to the probability of the answer **maybe** as the probability of failure of $S_{\mathcal{E}}$.

As long as the answer "**maybe** " is rare and the running time of $S_{\mathcal{E}}$ is fast, we will gain something by executing $S_{\mathcal{E}}$ first and then $T_{\mathcal{E}}$ only if necessary.

We will restrict ourselves to special classes of properties, namely hereditary properties in quasi orderings.

A **quasi ordering (qo)** $\mathcal{Q} = (\mathcal{V}, \preceq)$ is a set $\mathcal{V}$, the **objects**, together with a transitive and reflexive relation $\preceq$. Examples for this kind of structure are

- the set of subsets $\mathcal{V} = 2^V$ of a (possibly infinite) set V equipped with $\subseteq$,
- a class $\mathcal{V}$ of finite graphs equipped with
  - the induced subgraph relation $\underset{ind}{\preceq}$,
  - the weak subgraph relation $\underset{sub}{\preceq}$,
  - the graph minor relation $\underset{min}{\preceq}$,
- a class $\mathcal{V}$ of finite orders equipped with the induced suborder relation, also denoted with $\underset{ind}{\preceq}$.

A qo is called well founded if every descending chain $v_1 \succeq v_2 \cdots$ becomes stationary, i.e. from a certain index all v_i are equivalent. In the sequel all qo will be well founded. We may assume this for simplicity since all objects will be encoded in such a way that the encoding length, *length*(.), will decrease in any descending chain. We will denote this as the objects being **properly encoded.**

A property $\mathcal{E}$ of the objects in $\mathcal{V}$ is called **hereditary**, h.p., if $v \preceq w$ and $\mathcal{E}(w)$ implies that $\mathcal{E}(v)$ holds as well. Consider the properties "planarity" and "perfectness". Both are hereditary in $\mathcal{G}_{ind}$ but while "planarity" is also hereditary in $\mathcal{G}_{min}$ "perfectness" is not.

In a well founded qo every h.p. is uniquely characterized by the set of **minimal forbidden substructures** or **obstructions** denoted with $\text{Obstr}_{\mathcal{E}}$. For example the well known obstructions in $\mathcal{G}_{min}$ for "planarity" are $K_{3,3}$ and K_5. A set of obstructions need not be finite. For "perfectness" which is a h.p.in $\mathcal{G}_{ind}$ this set contains all odd holes and antiholes.

A distribution for a properly encoded qo $\mathcal{Q} = (\mathcal{V}, \preceq)$ is given by a family of distributions on the sets $\mathcal{V}_n = \{v \in \mathcal{V} \mid length(v) = n\}$. For example we may assume for each n that all $v \in \mathcal{V}_n$ are equally likely, i.e., that $\mathbf{Pr}(\{v\}) = 1/|\mathcal{V}_n|$ for all $v \in \mathcal{V}_n$. To avoid conflicts in notation we denote symbols coming from probability theory in **boldfont**; **Pr** for a probability measure, **E** for an expectation, and **p** and $\mathbf{q} = 1 - \mathbf{p}$ for probabilities. We will also abbreviate "random variable" to r.v.

The probability model for graphs we use will be the following. All graphs will have vertices labeled from 0 to $n-1$ and two isomorphic graphs with different labelings will be considered to be different. If $0 \leq p(n) \leq 1$ is some function we will assume a distribution on the set of graphs with n vertices such that the probability for a labeled edge being present in a randomly chosen graph will be $p(n)$. Compare the book of Bollobás [Bol85] for an introduction into the theory of random graphs. We will denote the set of all graphs equipped with this distribution with $\mathcal{G}^{\mathrm{p}}$. If in addition we assume that one of the relations mentioned above is given we denote this with the corresponding subscript, e.g. $\mathcal{G}^{\mathrm{p}}_{ind}$. For simplicity we denote the symmetric case $p(n) = 1/2$, in which every graph is equally likely, by $\mathcal{G}_{ind}$ and $\mathcal{G}_{sub}$.

3 Induced Subgraphs

In this section we illustrate our general principle of constructing fast expected time algorithms at the special case of hereditary properties in $\mathcal{G}_{ind}$. The algorithm we present here is rather simple. Its applicability, however, is limited to $\mathcal{G}_{ind}$ as its analysis is based on an estimation of the number of H-free graphs. In the next section we will explain a more general principle.

In the rest of this section $\mathcal{E}$ denotes an arbitrary but fixed hereditary property and H denotes a minimal obstruction for $\mathcal{E}$, i.e. a graph with minimal number of vertices which does not have the property $\mathcal{E}$. With $Forb_n(H)$ we denote the number of (labeled) graphs on n vertices without induced H-subgraph. Furthermore we assume that `exact-test`(G) is an algorithm which tests whether a given graph G has the property $\mathcal{E}$. We denote its worst-case running time on graphs with n vertices by $run_E(n)$. For any integer $k \leq |V(G)|$ let $G[0, k-1]$ denote the subgraph of G induced by the first k vertices.

Algorithm 1 `test-property`(G)

Input: *Graph G with n vertices, property $\mathcal{E}$, obstruction H;*
Output: **true** *if G has property $\mathcal{E}$,* **false** *otherwise.*

(1) $k := 1$;
(2) **while** $k \leq n$ **do begin**
(3) **if** $G[0, k-1]$ *contains* H **then return false**
(4) **else** $k := k+1$;
(5) **end**;
(6) **return** `exact-test`(G).

The correctness of the algorithm is immediate. To bound the average running time we first state two observations which are easily seen to be true.

(1) *If* `test-property` *stops with $k = l$, then $O(l^{|V(H)|})$ is an upper bound for the running time of* `test-property`.

(2) *The number of graphs on n vertices for which* `test-property` *completes at least l passes of the while-loop is exactly* $\mathrm{Forb}_l(H) \cdot 2^{\binom{n}{2}-\binom{l}{2}}$.

Let $ave_T(n)$ denote the average running time of the algorithm `test-property` on graphs with n vertices, assuming that every graph is equally likely. An upper bound for $ave_T(n)$ is then easily derived:

$$ave_T(n) = 2^{-\binom{n}{2}} \cdot \sum_{|V(G)|=n} run_T(G) \tag{3}$$

$$\leq 2^{-\binom{n}{2}} \left(\sum_{k=1}^{n} O\left(k^{|V(H)|}\right) Forb_k(H)\, 2^{\binom{n}{2}-\binom{k}{2}} + run_E(n)\, Forb_n(H) \right)$$

A consequence of the main result in [PS92b] is that for every graph H there exists a constant $c_H > 0$ such that

$$Forb_n(H) = 2^{(1-c_H)\binom{n}{2}+o(n^2)}. \tag{4}$$

This then immediately implies the following theorem.

Theorem 2. *Every property of graphs that is hereditary in $\mathcal{G}_{ind}$ and that has a test whose worst case running time is $2^{o(n^2)}$ can be tested in $O(1)$ expected time.* □

Corollary 3. *The following properties of graphs in $\mathcal{G}_{ind}$ can be tested in constant expected time: k-colorability, perfectness, monochromatic triangle, partition into cliques, partial k-tree.* □

For k-colorability Wilf [Wil84] had previously obtained such a constant expected time algorithm by a slightly more complicated argument. Here we prove the above corollary only for graph perfectness. The remaining properties follow similarly; compare [GJ79] for precise definitions of these problems.

A graph is called **perfect** if $\chi(G') = \omega(G')$ for all induced subgraphs G' of G. This problem is also chosen for an example, for which we give a more involved algorithm to achieve good expected time. This is so because we later we will need that the exponent of the running time is less than $n \log n$. To obtain a worst case complexity that suffices to apply Theorem 2 we would simply enumerate all subgraphs G' and calculate $\chi(G')$ and $\omega(G')$ for each of them, say. This leads to a worst case running time of $2^{O(n \log n)}$ which would be sufficient to prove Corollary 3.

For the more involved algorithm our basic idea is to use Lovász's Perfect Graph Theorem, namely that G being perfect is equivalent to

$$\omega(G') \cdot \alpha(G') \geq \left|V\left(G'\right)\right| \tag{5}$$

for all $G' \underset{ind}{\preceq} G$. See the book of Golumbic [Gol80] for more details and references.

Lemma 4. *Perfectness can be tested in running time $2^{O(n)}$.*

Algorithm 5 `test-perfectness`

Input: *Graph G with n vertices*
Output: **true** *if G is perfect,* **false** *otherwise.*

(1) **for** *all* $G' \preceq_{ind} G$ *with* ≤ 2 *vertices* **do** *initialize* $\omega[G']$ *and* $\alpha[G']$
(2) **for** $i := 3$ **to** n **do begin**
(3) **for** *all* $G' \preceq_{ind} G$ *with* i *vertices* **do begin**

(4) $$\alpha[G'] := \begin{cases} i & \text{if all } H \preceq_{ind} G' \text{ are stable, and} \\ \max_{H \preceq_{ind} G'} \alpha[H] & \text{otherwise} \end{cases}$$

(5) $$\omega[G'] := \begin{cases} i & \text{if all } H \preceq_{ind} G' \text{ are cliques, and} \\ \max_{H \preceq_{ind} G'} \omega[H] & \text{otherwise} \end{cases}$$

(6) **if** $\left(\omega[G'] \cdot \alpha[G'] < i\right)$ **then return false**
(7) **end**
(8) **end**
(9) **return true**

Proof. Consider Algorithm 5, `test-perfectness`, to test whether or not a graph G is perfect.

First we visit all induced subgraphs G' of G to calculate $\omega(G')$ and $\alpha(G')$ and then test it with help of Lovász's Perfect Graph Theorem for perfectness. This can be done in such a way that G' is only visited after all its induced subgraphs H have been visited before. If G' is a clique or independent set all its induced subgraphs H are so. This lets us detect whether or not G' has such a structure.

If it is not a clique or independent set then one of its induced subgraphs H contains a maximal clique or a maximal independent set respectively. So we just have to calculate the maximum over all such subgraphs.

Thus we calculate $\omega(G')$ and $\alpha(G')$ for all G' correctly and so we have correctness of the test for perfectness.

For the running time observe that the loop over the induced subgraphs G' gives the factor 2^n. The calculation of $\omega(G')$ and $\alpha(G')$ give $O(n)$ since we only have to consider those induced subgraphs H that have exactly one vertex less than G'. So in total we have $O(n \cdot 2^n) = O\left(2^{n+\log n}\right) = 2^{O(n)}$. □

4 Samples of Small Substructures

4.1 Samples

In this section we formalize the choice of several independent and small substructures of a large object in $\mathcal{V}$. For example we want to choose several independent subgraphs of a certain size from a graph.

A **sample** $\mathbf{X}$ of a qo $\mathcal{Q} = (\mathcal{V}, \preceq)$ is a family of r.v.'s $\mathbf{X}^i_{n,\mu}$: $\mathcal{V}_n \rightarrow \mathcal{V}_\mu$ with the following properties for all $\mu \in \mathbb{N}$:

(i) $\mathbf{X}^i_{n,\mu}(v) \preceq v$.
(ii) There is a non-decreasing function $l_\mu(.)$ such that for each $n \in \mathbb{N}$ the set of r.v.'s $\left\{\mathbf{X}^1_{n,\mu}, \ldots, \mathbf{X}^{l_\mu(n)}_{n,\mu}\right\}$ is independent.
(iii) There is a probability $0 < \mathbf{p}_\mu < 1$ such that for all $n \in \mathbb{N}$, $v_0 \in \mathcal{V}_\mu$ and all $0 < i \leq l_\mu(n)$ the r.v.'s $\mathbf{X}^i_{n,\mu}$ fulfill $\mathbf{Pr}(\mathbf{X}^i_{n,\mu} = v_0) \geq \mathbf{p}_\mu$.

We call l_μ the **sample length** and $\mathbf{p}_\mu$ the **sample probability** of $\mathbf{X}$. With $\mathbf{q}_\mu$ we denote $1 - \mathbf{p}_\mu$. $\mathbf{q}_\mu$ is an upper bound for $\mathbf{Pr}\big(\mathbf{X}^i_{n,\mu} \neq v_0\big)$. The following lemma estimates the probability that a certain obstruction v_0 appears as substructure of an element v.

Lemma 6. *Let* $\mathbf{X}$ *be a sample,* $v_0 \in \mathcal{V}_\mu$, $v \in \mathcal{V}_n$. *Then* $\mathbf{Pr}(v_0 \not\preceq v) \leq (\mathbf{q}_\mu)^{l_\mu(n)}$

Proof. This follows directly from the independence of the r.v.'s $\{\mathbf{X}^i_{n,\mu} \mid i \leq l_\mu(n)\}$ and from the fact that

$$\mathbf{Pr}\left(v_0 \not\preceq v\right) \leq \mathbf{Pr}\left(\bigwedge_{i \leq l_\mu(n)} v_0 \neq \mathbf{X}^i_{n,\mu}(v)\right) \leq \prod_{i=1}^{l_\mu(n)} \mathbf{Pr}\left(v_0 \neq \mathbf{X}^i_{n,\mu}(v)\right) \tag{6}$$

□

We call a property $\mathcal{E}$ on $\mathcal{Q}$ **sparse** if $\lim_{n\to\infty} \mathbf{Pr}(\mathcal{E}\,(v) = \mathbf{true} \mid v \in \mathcal{V}_n) = 0$,

$$\lim_{n\to\infty} \frac{|\{v \in \mathcal{V}_n \mid \mathcal{E}\,(v) = \mathbf{true}\,\}|}{|\mathcal{V}_n|} = 0. \tag{7}$$

It is **dense** if this limit tends to 1. The following theorem shows that h.p.'s are sparse in a very general setting.

Theorem 7. *Let* $\mathcal{Q} = (\mathcal{V}, \preceq)$ *be properly encoded with sample* $\mathbf{X}$ *such that the sample length* $l_\mu(.)$ *is unbounded for every* μ. *Then every non-trivial h.p. in* $\mathcal{Q}$ *is sparse, that is satisfies* $\lim_{n\to\infty} \mathbf{Pr}(\mathcal{E}\,(v) = \mathbf{true} \mid v \in \mathcal{V}_n) = 0$.

Proof. Let $\mathcal{E}$ be a h.p.. Since it is non-trivial there is some $v_0 \in \mathcal{V}_\mu$ for some μ such that $\neg\mathcal{E}\,(v_0)$ holds. Then for all $v \in \mathcal{V}_n$ with $\mathcal{E}\,(v)$ we have that v_0 is not below v, $v_0 \not\preceq v$. So

$$\mathbf{Pr}(\mathcal{E}\,(v) = \mathbf{true}\,) \leq \mathbf{Pr}(v_0 \not\preceq v) \leq (\mathbf{q}_\mu)^{l_\mu(n)} \tag{8}$$

Since $l_\mu(.)$ is unbounded this shows the claim. □

A **sample algorithm** A_X is an algorithm that incrementally produces a sample $\mathbf{X}$. We assume that such an algorithm is implemented as two distinct subroutines. The first one performs some initialization and the other one is given in such a way that for all $0 < i \leq l_\mu(n)$ the i-th call of this routine outputs $\mathbf{X}^i_{n,\mu}(v)$. We denote with $t^{init}_{\mu,A}$ and $t^{inc}_{\mu,A}$ the time such an algorithm needs for an initial phase and for each incremental step, respectively.

Now let $\mathcal{E}$ be a h.p.and $v_0 \in \mathcal{V}_\mu$ be an obstruction for $\mathcal{E}$, i.e., $\mathcal{E}$ is false on v_0. In addition assume we are given an algorithm $T_\mathcal{E}$ that outputs $\mathcal{E}(v)$ with running time $t_{T_\mathcal{E}}(n)$. Consider the known test routine $T_\mathcal{E}$ as being expensive; $t_{T_\mathcal{E}}(n)$ grows faster than we want. Here "growing fast" can mean different things:

- super-polynomial,
- linear (or low polynomial) with enormous constants of proportionality or
- super-polylogarithmic on a parallel machine, say,

depending on the setting we want to deal with.

The following Algorithm **ave**, implements a strategy which tries to avoid the call to the exact, but expensive algorithm $T_\mathcal{E}$. It simply tests whether or not $v_0 = \mathbf{X}^i_{n,\mu}(v)$ for some i before it calls $T_\mathcal{E}$.

Algorithm 8 $\quad \mathbf{ave}_{A_\mathrm{X},T_\mathcal{E},v_0}(v)$

Input: $v \in \mathcal{V}$ *and* $n = \text{length}(v)$
Output: $\mathcal{E}(v)$
(1) *Initialize* A_X *with* v *and* $n = length(v)$;
(2) **for** $i := 1$ **to** $l_\mu(n)$ **do begin**
(3) $\quad \mathbf{X}_i := A_\mathrm{X}$;
(4) $\quad$ **if** $(\mathbf{X}_i = v_0)$ **then return false** ;
(5) $\quad$ **end** ;
(6) **return** $T_\mathcal{E}(v)$.

Lemma 9. *For fixed* v_0 *the average running time of* $\mathbf{ave}_{A_\mathrm{X},T_\mathcal{E},v_0}$ *is bounded by*

$$O\Big(t^{init}_{\mu,A}(n) + t^{inc}_{\mu,A}(n) + \mathbf{q}_\mu^{l_\mu(n)} \cdot t_{T_\mathcal{E}}(n)\Big). \tag{9}$$

Proof. The first term is obvious. The third term is just an upper bound on the probability that $T_\mathcal{E}$ is executed multiplied with its running time. For the second term observe that the probability for the i-th execution of the for loop is bounded by $\mathbf{q}_\mu^{i-1}$. So the average time this loop needs is at most

$$\sum_{i=1}^{l_\mu(n)} \mathbf{q}_\mu^{i-1} \cdot t^{inc}_{\mu,A}(n) = t^{inc}_{\mu,A}(n) \cdot \sum_{i=1}^{l_\mu(n)} \mathbf{q}_\mu^{i-1} \le t^{inc}_{\mu,A}(n)/\mathbf{p}_\mu. \tag{10}$$

But $\mathbf{p}_\mu$ is a constant depending only on v_0 and not on v. □

We conclude with the following theorem:

Theorem 10. *Let* $\mathcal{E}$ *be a h.p.on* $\mathcal{Q} = (\mathcal{V}, \preceq)$ *and* A_X *be a sample algorithm for* $\mathcal{Q}$. *Suppose there is an algorithm* $T_\mathcal{E}$ *to test* $\mathcal{E}$ *that has worst case running time* $t_T(n) = O(\mathbf{q}_\mu{}^{-l_\mu(n)})$ *for all* μ. *Then* $\mathcal{E}$ *can be tested in average time* $O(t^{init}_{\mu,A}(n) + t^{inc}_{\mu,A}(n))$.

Proof. We may assume that $\mathcal{E}$ is non-trivial, i.e., it has an obstruction $v_0 \in \mathcal{V}_\mu$ for some μ. Consider the running time of $\mathbf{ave}_{A_X, T_\mathcal{E}, v_0}$. The third term of the complexity given in Lemma 9 is $O\left(\mathbf{q}_\mu^{l_\mu(n)} \cdot t_T(n)\right) \leq O(1)$. □

4.2 Induced Suborders

We exemplify this approach on some classes of partial orders. Several models for random graph orders have been studied in the literature. The two that perhaps are the most common ones are the uniform distribution on all orders and one introduced by Winkler [Win85]. The first is analogous to the model used above for graphs. In contrary to the models for graphs, however, it doesn't allow independent choices of suborders. So our approach fails for this model and we restrict our attention to the second model introduced by Winkler. It assigns a certain probability distribution to all orders of dimension at most k for some fixed value k.

Both models differ drastically in the behavior random orders have; e.g. for the first a general 0-1-law for first-order-logic properties applies, cf. [Com88], whereas for the second model Winkler, cf. [Win89], gave examples for such properties with limit probability $\neq 0, 1$ for $k = 3$. Also standard parameters of orders as e.g. width and height differ on expectation in the two models.

To briefly define the concepts let $P = (V, <)$ be a partially ordered set, poset or order for short. Let $L_1, \ldots, L_k$ be a set of linear extensions of P. By $\bigcap_{i=1}^k L_i$ we denote the order on V where two elements in V are related iff they are related, in the same direction, in all the L_i. Clearly we have $P \underset{sub}{\preceq} \bigcap_{i=1}^k L_i$. If we also have equality then $L_1, \ldots, L_k$ is said to realize P. If k is minimal such that there exists a realizer $L_1, \ldots, L_k$ for P, P is said to have **dimension** k.

The r.v. $\mathbf{P}_k(n)$ is given by independently choosing k linear orders $L_1, \ldots, L_k$ over $V = \langle n \rangle = \{0, \ldots, n-1\}$ and taking their intersection $\bigcap_{i=1}^k L_i$. By $\mathcal{P}^k(n)$ we denote the corresponding probability space on the set of orders on $\langle n \rangle$ of dimension at most k. Observe that every fixed such order P appears with probability at least $\frac{1}{(n!)^k}$ in $\mathcal{P}^k(n)$. Furthermore let $\mathcal{P}^k = \bigcup_n \mathcal{P}^k(n)$ and let $\mathcal{P}_{ind}^k$ denote $\mathcal{P}^k$ equipped with $\underset{ind}{\preceq}$.

From Winkler's work, see [Win85], we easily deduce the following remarks:

Remark 1 *Let $V^1 \subset \langle n \rangle$, $n_1 = |V^1|$ and $\mathbf{P}_k(n)\big|_{V^1}$ be the r.v. obtained by restricting $\mathbf{P}_k(n)$ to V^1. Then $\mathbf{P}_k(n)\big|_{V^1}$ has the same distribution as $\mathbf{P}_k(n_1)$.*

Remark 2 *Let $\{V^1, \ldots, V^r\}$ be a family of mutually disjoint subsets of $\langle n \rangle$ then the set of r.v.'s $\{\mathbf{P}_k(n)\big|_{V^1}, \ldots, \mathbf{P}_k(n)\big|_{V^r}\}$ is independent.*

With Remark 1 we may identify $\mathbf{P}_k(n)\big|_{V^1}$ and $\mathbf{P}_k(n_1)$ by renaming the elements in V^1 accordingly. With Remark 2 we may assign to each appropriate subset $\{V^1, \ldots, V^r\}$ an independent set of r.v.'s $\{\mathbf{X}^i \mid i = 1, \ldots, r\}$, each $\mathbf{X}^i$ living on $\mathcal{P}^k(n_i)$.

This allows us to define a sample on $\mathcal{P}_{ind}^k$. For $\mu \leq n$ let $l_\mu(n) = \lfloor n/\mu \rfloor$ and for $i = 0, \ldots, l_\mu(n) - 1$ let $V_{n,\mu}^i = \{i\mu, \ldots, (i+1)\mu - 1\}$. Now let $\mathbf{X} = \{\mathbf{X}_{n,\mu}^i\}$ denote the corresponding set of r.v.'s. With what is said above $\mathbf{X}$ is a sample. Thus with Theorem 7 we get:

Corollary 11. *Every hereditary property in $\mathcal{P}^k_{ind}$ is sparse.*

Now let $A_{\mathbf{X}}$ denote the trivial algorithm that produces $\mathbf{X}$ incrementally. If we assume to have random access to the order that is given as input Theorem 10 gives us:

Corollary 12. *Every hereditary property in $\mathcal{P}^k_{ind}$ that has a test in $2^{o(n)}$ time in the worst case can be tested on average in $O(1)$ time.*

4.3 A Prototype for a Sample Algorithm

Now we give a prototype of a sample algorithm that will be useful for different combinatorial structures. Let us assume that the structure is given as a set S of k-element subsets of $\langle n\rangle = \{0,\ldots,n-1\}$, i.e. $S \subseteq \binom{\langle n\rangle}{k}$. k is assumed to be a fixed constant, e.g 2 for graphs or 3 for 3-SAT formulas. We denote this set of sets by $\mathcal{S}_k(n)$, i.e. $\mathcal{S}_k(n) = 2^{\binom{\langle n\rangle}{k}}$.

We assume that $\mathbf{Pr}(s \in S) = \mathbf{p} > 0$ for all $s \in \binom{\langle n\rangle}{k}$ and some value $\mathbf{p}$. All events $s \in S$ occur independently of each other. If we denote $1-\mathbf{p}$ by $\mathbf{q}$ any given S appears with a probability of $\mathbf{r}_S = \mathbf{p}^{|S|}\mathbf{q}^{\binom{n}{k}-|S|}$. We denote the corresponding probability space with $\mathcal{S}^{\mathbf{p}}_k(n)$ and with $\mathcal{S}^{\mathbf{p}}_k = \bigcup_n \mathcal{S}^{\mathbf{p}}_k(n)$.

An induced substructure $S\big|_I \subseteq \langle\mu\rangle$ of S for some $\mu < n$ is obtained by a subset $I = \{i_0 < i_1 < \cdots < i_{\mu-1}\} \subseteq \langle n\rangle$ by

$$\{j_1,\ldots,j_k\} \in S\big|_I \iff \{i_{j_1},\ldots,i_{j_k}\} \in S. \tag{11}$$

That is we take all $s \in S \cap \binom{I}{k}$ and rename their elements such that the resulting sets are in $\binom{\langle\mu\rangle}{k}$.

For graphs this means for example that we distinguish induced subgraphs for which the elements are numbered in different order.

Now it is clear that in order to produce samples of substructures of S it is sufficient to produce "independent" subsets of $\langle n\rangle$. Consider the following algorithm that produces μ element subsets of $\langle\mu^d\rangle$.

Algorithm 13 `sample`(μ, d, a)

Input: *Positive integers μ, d and a; $n = \mu^d$.*
Output: *A sequence $(V_1,\ldots,V_{l_\mu(n)})$ of μ element subsets of $\{a,\ldots,a+n-1\}$.*
(1) $m := n/\mu$
(2) **if** $(d > 1)$ **then for** $i := 0$ **to** $\mu-1$ **do** `sample`$(\mu, d-1, a+i\cdot m)$
(3) **for** $i := 0$ **to** $m-1$ **do** *Output* $\{a+i, a+i+m,\ldots,a+i+(\mu-1)m\}$

Lemma 14. *For every call of* `sample`$(\mu, d, 0)$ *with* $\mu, d \geq 2$ *we have:*

(i) Every pair $i \neq j$ *occurs at most once in an output set.*
(ii) Every subset $V_0 \subseteq \langle \mu^d \rangle$ *with* $|V_0| > 1$ *occurs at most once as a subset of an output set.*
(iii) $d\mu^{d-1}$ *pairwise distinct subsets of* $\langle \mu^d \rangle$ *are produced.*

Proof. "(ii)" follows directly from "(i)". To see "(i)" we first restrict ourselves to the sets given in loop (3). For them this is clear since every single point appears in exactly one subset produced there. But all pairs $i \neq j$ that are used here end up in different recursive calls. So they will never appear together in one of the subsets generated by the recursive calls.

To see (iii) apply induction on d. □

It is easy to see that the following iterative variant produces exactly the same output:

Algorithm 15 `sample`(μ, d, a)

Input: *Positive integers* μ, d *and* a; $n = \mu^d$.
Output: *A sequence* $(V_1, \ldots, V_{l_\mu(n)})$ *of* μ *element subsets of* $\{a, \ldots, a + n - 1\}$.

(1) **for** $c := 0$ **to** $d - 1$ **do begin**
(2) $r := \mu^c$
(3) **for** $i_0 := 0$ **to** $n - 1$ **step** $\mu{\cdot}r$ **do begin**
(4) **for** $i := i_0$ **to** $i_0 + r - 1$ **do**
(5) *Output* $\{i, i + r, \ldots, i + (\mu - 1)r\}$
(6) **end**
(7) **end**

Theorem 16. *Let* k, $\mathbf{p}$, μ *and* $S_0 \in \mathcal{S}_k(\mu)$ *be fixed and* $\mathbf{r} = \mathbf{r}_{S_0}$. *Then there is an algorithm that finds* S_0 *as substructure in a given input* $S \in \mathcal{S}_k^{\mathbf{p}}$ *with expected running time of* $O(1)$ *and has a probability of failure of* $(1 - \mathbf{r})^{\Theta(n \log n)}$.

Proof. Lemma 14 (iii) guarantees that `sample`$(\mu, d, 0)$ for μ and d such that $\mu^{d-1} < n \leq \mu^d$ produces $d\mu^{d-1} = \Theta\left(\lfloor \log_\mu n \rfloor \mu^{\lfloor \log_\mu n \rfloor - 1}\right) = \Theta(n \log n)$ different subsets. By (ii) of the same lemma we know that all substructures produced by restricting S to these subsets appear independently of each other.

For the running time observe that the iterative variant of `sample` only needs constant time for initialization and also for incrementally producing the each new sample. □

The following slight generalization of Theorem 2 is an easy application for $k = 2$.

Theorem 17. *Every property of graphs that is hereditary in* $\mathcal{G}_{ind}^{\mathbf{p}}$, $\mathbf{p}$ *a fixed constant, and that has a test whose worst case running time is* $2^{o(n \log n)}$ *time can be tested in* $O(1)$ *expected time.* □

Corollary 18. *The following properties of graphs in $\mathcal{G}^{\mathbf{p}}_{ind}$ can be tested in constant expected time: k-colorability, perfectness, monochromatic triangle, partition into cliques, partial k-tree.* □

5 Satisfiability

In this section we give an application of another kind of combinatorial structure, namely CNF-formulas. We assume that such formulas are given over the set of variables $\{v_0, v_1, \ldots\}$ and we always write clauses with the literals in order, e.g. $(v_2 \vee \overline{v_5} \vee \overline{v_{97}})$ might be such a clause and $(\overline{v_{45}} \vee \overline{v_{46}} \vee v_{52}) \wedge (v_5 \vee v_6 \vee v_9)$ might be such a formula. We will also assume that the number of literals that appear in a clause is exactly k for some constant k, that the variables used in a specific formula F is exactly $\{v_0, v_1, \ldots, v_{n(F)}\}$ for some value $n(F)$, and that no clause is repeated. We denote this class of formulas with $\mathcal{F}_k$. This class is equipped with a probability measure such that the probability that a certain clause appears in a formula over n variables is $\mathbf{p}$, and such that all clauses appear independently. We denote this probability space with $\mathcal{F}^{\mathbf{p}}_k$.

We may then assign to each clause a set in $\binom{\langle \mathbf{N}_0 \rangle}{k}$ and to each formula a system $S(F)$ of sets in $\binom{\langle n(F) \rangle}{k}$. Since every such set can be obtained from different clauses we also have to administrate a label $\lambda(s) \subseteq \langle 2 \rangle^k$ for each $s \in S(F)$. Every clause in F contributes a k-vector of 0's and 1's to the corresponding $s \in S(F)$. We interpret a 1 as the variable being negated and a 0 as being not negated, e.g. $(v_2 \vee \overline{v_5} \vee \overline{v_{97}})$ gives the vector $(0, 1, 1)$.

By $S^\lambda(F)$ we denote $S(F)$ together with the associated sets of labels of all $s \in S(F)$. So for every formula F the system $S^\lambda(F)$ is a set of elements of $\binom{\langle n(F) \rangle}{k} \times 2^{\langle 2 \rangle^k}$. With this encoding we represent every formula in a unique way.

This assignment of sets with labelings for each formula defines a probability measure on $2^{\binom{\langle n(F) \rangle}{k} \times 2^{\langle 2 \rangle^k}}$. It has the property that given a k-set s in $\binom{\langle n(F) \rangle}{k}$ the probability that λ_0 appears as the label set of s is $\mathbf{p}^{|\lambda_0|}(1 - \mathbf{p})^{2^k - |\lambda_0|}$. It is thus bounded from below by some constant α. In particular the case that a certain set s does not appear can also be estimated by this value.

Theorem 19. *Let k and $\mathbf{p}$ be fixed. Then there is an algorithm that decides every satisfiability formula $F \in \mathcal{F}^{\mathbf{p}}_k$ in constant expected time.*

Proof. Observe first that satisfiability is a h.p. with respect to "taking subformulas" since a truth assignment of the whole formula satisfies every subformula. It is clear that this extends to the set systems with label $S^\lambda(F)$ as given above. Given any fixed unsatisfiable formula F_0 our algorithm $\texttt{sample}_{k,n(F_0)}$ finds $S^\lambda(F_0)$ with high probability in $S^\lambda(F)$ for any random input F. We may even just check our subformula of size μ if it is satisfiable or not. Theorem 16 shows that the probability of failure vanishes with exponent $\Omega(n(F) \log n(F))$. Hence, if we apply the brute forward satisfiability test to F, its running time $O(n(F)^k 2^{n(F)}) = 2^{\mathcal{O}(n(F))}$ is cancelled out by this probability. □

6 Classes of Sparse Graphs

In this section we generalize the results obtained so far to distributions on graphs specified by various vanishing edge probability functions $p = p(n)$.

6.1 Subgraphs

Let us recall the general approach of Section 3. The success of the algorithm `test-property` mainly relied on the fact that the number of graphs containing no induced H-subgraph is exponentially small compared to the number of all graphs. If on the other hand one considers properties in $\mathcal{G}^p_{ind}$ for arbitrary edge probability functions $p = p(n)$ one would need to know the ratio between H-free subgraphs with a given number of edges and the number of all graphs with that number of edges. While this is in general not known for *induced* subgraphs the following result of [JŁR90] provides such bounds for excluded weak subgraphs.

Theorem 20. *Let $G_{n,p}$ be a random graph with edge probability $p = p(n)$ and let E_H denote the event that $G_{n,p}$ contains no (weak) H-subgraph. Then there exists a constant $c_H > 0$ such that*

$$\Pr(E_H) \leq exp(-c_H \min\left\{n^{|A|}p^{e(A)} \mid A \subseteq H, e(A) > 0\right\}). \tag{12}$$

Similar as in Section 3 we therefore obtain the following theorem.

Theorem 21. *Every property of graphs that is hereditary in $\mathcal{G}^p_{sub}$ and that has a test whose worst case running time is bounded by $2^{f(n)}$ and that has an obstruction H such that $f(n) \ll \min\{n^{|A|}p^{e(A)} \mid A \subseteq H, e(A) > 0\}$ can be tested in $O(1)$ expected time.*

In order to apply this theorem one needs good exact algorithms to achieve fast expected time algorithm for small edge probabilities p. In other words, the theorem provides us with a tradeoff between the worst case complexity and the edge probability p for which simple and fast expected time algorithms can be constructed.

6.2 Minors

Because of a theorem of Thomason, see below, hereditary properties w.r.t. $\preceq_{min}$ are only interesting for graphs with at most a linear number of edges. So let $G = (V, E)$ be a graph on n vertices with cn edges for some fixed constant c. Then the situation around any vertex looks like a tree (or nearly a tree) as the probability that two vertices have a common neighbor is small. If we contract this tree we obtain a new graph with slightly higher vertex degrees.

This is the main idea behind the Algorithm `contract`. Let k be a fixed number, choose $\frac{k-1}{k}n$ edges in G and contract them. We end up with a graph G' that has about than $\frac{n}{k}$ vertices. If we allow loops and multiple edges it would have $\frac{kc-k+1}{k}n$ edges. So it has about $kc - k + 1$ edges per vertex.

So this is a way to *concentrate* a graph in a certain sense. There are two problems left, first the number of cycles we generate with the chosen edges, and second how many loops and multiple edges do appear.

Algorithm 22 contract (G, k)

Input: *Graph* $G = (V, E)$, $n = |V|$, $|E| \geq |V|$ *and integer* k
Output: *Graph* G'
(1) *Choose a set* $E_0 \subseteq E$ *with* $|E_0| = \frac{k-1}{k}|V|$ *at random*
(2) *Let* V' *be the components of* $G\Big|_{E_0}$.
(3) **for all** $e = (v, w) \in E$ **do** *add* (v', w') *to* E'
(4) **return** (V', E')

Theorem 23. *For* $c > 1$ *and* $s > 0$ *let* $\mathcal{G}_{min}^{c/n,s} \subseteq \mathcal{G}_{min}^{c/n}$ *be the subclass of graphs that have* K_s *as a minor.*

(i) There is an algorithm running in linear time that outputs either **maybe** *or an evidence for* K_s *being a minor.*
(ii) The probability that the algorithm gives the output **maybe** *goes exponential to* 0 *if the size of the input graph grows.*

Proof. According to a theorem of Thomason [Tho84] for every sufficiently large s there is a constant d_s such that all graphs $G = (V, E)$ with $|E| > d_s|V|$ contain a K_s as a minor.

If $c > 2d_s$ we know that the number of edges of a graph in $\mathcal{G}_{ind}^{c/n}$ is larger than $d_s n$ with high probability. So there is nothing to show. Let us therefore assume that $c \leq d_s$. We first consider the case $c > 2$. Here we only have to show that for an appropriate choice of k the output of contract has more than $d_s|V'|$ many edges with high probability.

This can easily be seen by regarding contract as a r.v.. If we consider a typical element in the image we see that it has more than $\frac{n}{k}$ vertices and the probabilities of the edges are independent. There are $\frac{n^2}{2k^2}$ possible new edges (including loops) and any of these appears with probability $\frac{k(c-2)+2}{2k} \cdot n \cdot \frac{2k^2}{n^2} = \frac{k^2(c-2)+2k}{n}$. So if we have chosen k large enough the expected number of remaining edges is larger than Thomason's constant.

All this can be done in linear time. First we compute the components, then we rearrange the remaining edges of G such that loops and multiple edges disappear. This is achieved in linear time, if we use e.g. bucketsort for rearrangement of the edges.

Secondly, we consider the case that $1 < c < 2$. If the graph has more that $(\frac{1}{2}+\varepsilon)n$ edges, its largest component has with high probability cardinality at least $c_\varepsilon n$ for some constant c_ε depending on ε. This component has $c_\varepsilon n + \delta_\varepsilon n$ many edges, for some constant $\delta_\varepsilon > 0$. So we may regard "largest component" as a r.v. on the **connected** graphs on $c_\varepsilon n$ vertices and $(c_\varepsilon + \delta_\varepsilon)n$ edges. While the edges in such a graph are not fully independent, they are independent to a high extend. So we succeed with algorithm contract as seen above. □

As a consequence we get:

Theorem 24. *Every h.p. in $\mathcal{G}_{min}^{c/n}$ for $c \neq 1$ can be tested in linear expected time.*

Proof. Let s be such that all minimal obstructions for the property have less than s vertices. So every such obstruction is a subgraph of K_s. If $c > 1$, the theorem follows immediately with `contract` and the fact that every h.p. in $\mathcal{G}_{min}$ can be tested in cubic time, see e.g. [RS85b]. If on the other hand $c < 1$, we first observe that the probability that there exists a component such that more than 5 edges must be removed to make it acyclic is bounded by $O(n^{-4+\delta})$ (cf. [ER60]). Recall that the treewidth of a graph is bounded by some absolute constant, if every component of the graph has this property. So we may run Bodlaenders tree decomposition algorithm [Bod93] on the graph to get a constant sized decomposition in linear time. Now every h.p. can be tested in linear time (cf. [ALS91]). □

Observe also that the constant of proportionality for the algorithm depends "only" exponentially on the size of a minimal obstruction. So it might be worth executing our algorithm even when a worst case linear algorithm is known. This is e.g. the case for all properties excluding a planar graph, see [Bod93], where the constants are enormous compared to the size of a minimal obstruction.

References

[ALS91] Stefan Arnborg, Jens Lagergren, and Detlef Seese, *Easy problems for tree-decomposable graphs*, J. Algorithms **12** (1991), 308–340.

[BFF85] B. Bollobás, T.I. Fenner, and A.M. Frieze, *An algorithm for finding hamilton paths and cycles in random graphs*, in [STO85] (1985), 430–439.

[BK79] L. Babai and L. Kucera, *Canonical labelling of graphs in linear time*, in [FOC79] (1979), 39–46.

[Bod93] Hans L. Bodlaender, *A linear time algorithm for finding tree-decompositions of small treewidth*, in [STO93] (1993).

[Bol85] Belá Bollobás, *Random graphs*, Academic Press, London, 1985.

[Com88] K. J. Compton, *The computational complexity of asymptotic problems. I: Partial orders*, Inform. and Complexity **78** (1988), 103–123.

[DF89] M.E. Dyer and A.M. Frieze, *The solution of some random NP-hard problems in polynomial expected time*, J. Algorithms **10** (1989), 451–489.

[ER60] P. Erdős and A. Rényi, *On the evolution of random graphs*, Madyar Tnd. Akad. Mat. Kut. Int. Kőzl. **6** (1960), 17–61.

[Fel89] Michael R. Fellows, *The Robertson-Seymour theorems: a survey of applications*, see [GA89], 1989, pp. 1–18.

[FL85] Michael R. Fellows and Michael A. Langston, *Nonconstructive advances in polynomial-time complexity*, Inform. Process. Lett. (1985).

[FL88] Michael R. Fellows and Michael A. Langston, *Nonconstructive tools for proving polynomial-time decidability*, J. Assoc. Comput. Mach. **35** (1988), no. 3, 727–739.

[FL92] Michael R. Fellows and Michael A. Langston, *On well-partial-order theory and its application to combinatorial problems of VLSI design*, SIAM J. Disc. Math. **5** (1992), no. 1, 117–126.

[FOC79] 20th Annual Symposion On Foundations of Computer Science, IEEE, The Institute of Electrical and Electronics Engineers, IEEE Computer Society Press, 1979.

[FP83] John Franco and Marvin Paull, *Probabilistic analysis of the Davis Putnam procedure for solving the satisfiability problem*, Discrete Appl. Math. **5** (1983), 77–87.

[Fri90] A.M. Frieze, *Probabilistic analysis of graph algorithms*, Computing **7** (1990), 209–233.

[GA89] Graphs and Algorithms (R. B. Richter, ed.), American Math. Soc., Contemp. Math., 89, 1989, Proceedings of the AMS-IMS-SIAM joint Summer Research Conference 1987.

[GJ79] Michael R. Garey and David S. Johnson, Computers and Intractability, W. H. Freeman and Company, New York, 1979.

[Gol80] Martin C. Golumbic, *Algorithmic graph theory and perfect graphs*, Academic Press, London, New York, 1980.

[GPB82] A. Goldberg, P. Purdom, and C. Brown, *Average time analysis of simplified Davis-Putnam procedures*, Inform. Process. Lett. **15** (1982), 72–75, Corrigendum, *Inform. Process. Lett.* **16** (1982), 213.

[Gus92] Jens Gustedt, *Algorithmic aspects of ordered structures*, Ph.D. thesis, Technische Universität Berlin, 1992.

[HTL91] T. H. Hu, C. Y. Tang, and R. C. T. Lee, *An average case analysis of Monien and Speckmeyer's mechanical theorem proving algorithm*, in [ISA91] (1991), 116–126.

[ISA91] ISA '91 Algorithms (Wen-Lian Hsu and R.C.T. Lee, eds.), Springer-Verlag, 1991, LNCS 557, Proceedings of the 2nd International Symposion on Algorithms, Taipei, Republic of China.

[Iwa89] Kazuo Iwama, *CNF satisfiability test by counting and polynomial average time*, SIAM J. Comput. **18** (1989), no. 2, 385–391.

[JŁR90] Svante Janson, Thomasz Łuczak, and Andrzej Ruciński, *An exponential bound for the probability of nonexistence of a specified subgraph in a random graph*, Random Graphs '87 (M. Karoński, J. Jaworski, and A. Ruciński, eds.), John Wiley & Sons, New York, 1990, pp. 73–87.

[Joh84] David S. Johnson, *The NP-completeness column: An ongoing guide, "Solving NP-hard Problems Quickly (On Average)"*, J. Algorithms **5** (1984), 284–299.

[Möh90] Rolf H. Möhring, *Graph problems related to gate matrix layout and PLA folding*, Computational Graph Theory (Wien) (G. Tinhofer et al., eds.), Springer-Verlag, Wien, 1990, pp. 17–52.

[PS92a] Hans Jürgen Prömel and Angelika Steger, *Coloring clique-free graphs in linear expected time*, Random Structures and Algorithms **3** (1992), 374–402.

[PS92b] Hans Jürgen Prömel and Angelika Steger, *Excluding induced subgraphs III: A general asymtotic*, Random Structures and Algorithms **3** (1992), no. 1, 19–31.

[RS85a] Neil Robertson and Paul Seymour, *Disjoint paths – a survey*, SIAM J. Algebraic Discrete Methods **6** (1985), no. 2, 300–305.

[RS85b] Neil Robertson and Paul Seymour, *Graph minors – a survey*, Surveys in Combinatorics (Glasgow 1985) (I. Anderson, ed.), Cambridge Univ. Press, Cambridge-New York, 1985, pp. 153–171.

[STO85] Proceedings of the Seventeenth Anual ACM Symposion on Theory of Computing, ACM, Assoc. for Comp. Machinery, 1985.

[STO93] Proceedings of the Twenty Fifth Anual ACM Symposion on Theory of Computing, ACM, Assoc. for Comp. Machinery, 1993.

[Tho84] A.G. Thomason, *An extremal function for contractions of graphs*, Math. Proc. Cambridge Philos. Soc. **95** (1984), 261–265.

[Wil84] H. S. Wilf, *Backtrack: an $O(1)$ expected time algorithm for the graph coloring problem*, Inform. Process. Lett. **18** (1984), 119–121.

[Win85] Peter Winkler, *Random orders*, Order **1** (1985), 317–331.

[Win89] Peter Winkler, *A counterexample in the theory of random orders*, Order **5** (1989), 363–368.

Orders, k-sets and Fast Halfplane Search on Paged Memory*

Paolo G. Franciosa and Maurizio Talamo

Dipartimento di Informatica e Sistemistica, University of Rome "La Sapienza", via Salaria 113, I-00198 Rome, Italy. E-mail: {pgf,talamo}@dis.uniroma1.it

Abstract. We investigate the properties of the poset defined by the set inclusion relation on the set of possible answers to an halfplane search problem. We use these properties to design a static data structure for n points stored in pages of size b requiring $O(\frac{n\sqrt{n}}{b})$ worst case pages that reports a solution of size k by accessing $O(\log_b n \log\log n + \frac{k}{b})$ pages. This is the first sub-quadratic data structure for secondary storage with output sensitive (in the total number of accessed pages) query time, and exploits an original point clustering technique.

1 Introduction

Representing properties of a set of geometric objects using some partial order relation among them is a suggestive idea, and has been often exploited in a number of applications. Particularly this approach seems to be useful when we want to design an implicit data structure representing these properties. In this case the partial order relation encodes the relations defined among elements in the data structure.

An implicit data structure is necessary when we refer to computational models based on paged memories. The computational model underlying the concept of paged memory consists of a memory organized in pages of size b; the space complexity is measured by the number of pages used, and the query time is measured by the number of pages accessed in order to answer a query. This computational model is used to represent typical secondary memory management problems, and is becoming more and more representative since many applications work on geometric data stored on secondary memory.

In order to design an implicit data structure on paged memory starting from a partial order we can cluster the partial order, trying to reduce the number of pages accessed while answering a query. If the solution is found by navigating in the partial order, the clustering process should reduce the number of edges of the Hasse diagram of the poset between elements in different pages.

In this paper we study the halfplane search problem, and we show how, using a partial order relation, it is possible to design an output sensitive sub-quadratic

* Work partially supported by the ESPRIT II Basic Research Action Program of the European Community under contracts No.6881 "Algorithms, Models, User and Service Interfaces for Geography" and No.7141 "Algorithms and Complexity II"

data structure on a paged memory computational model. The *halfspace search* problem is classically stated as follows: given a set of n points $P \subseteq \mathbb{R}^d$, build a data structure that allows to efficiently determine which points lie in a closed query halfspace.

Many simple problems in computational geometry have been studied on paged memory [13]. On the other hand the halfspace search problem is one of the challenging problems in computational geometry; indeed the halfspace search problem is particularly interesting since it represents a typical "non-isothetic" query on geometric data, and it is a basic primitive for many complex geometric operations. This "non-isothetic"-ness characteristic critically affects the design of a paged data structure, leading to solutions whose performances are very poor, with respect to either space or query time, while optimal solutions are available on the classical RAM model. An optimal output-sensitive solution has been given in [4], in which a linear size static) solution is proposed that solves an halfplane search query in $O(\log n + k)$ worst case time, where k is the number of points retrieved, on the RAM computational model.

We show that it is possible to cluster the solutions to halfplane searches in order to obtain a data structure for secondary storage that requires $O\left(\frac{n\sqrt{n}}{b}\right)$ worst case pages and retrieves a solution of size k by accessing $O(\log_b n \log\log n + \frac{k}{b})$ pages. We reach our results by defining a partial ordering on the set of solutions by the set inclusion relation, and by showing that it is a lattice with linear dimension 2.

Note that up to now no output sensitive solutions have been found using a sub-quadratic number of pages.

The paper is structured as follows: in Sect. 2 some basic definitions on partial orders are given; the halfplane search problem is introduced in Sect. 3, where Subsect. 3.1 shows its connections with the study of circular sequences, while Subsect. 3.2 introduces the main results on the combinatorial structure of the set of answers to halfplane queries. Sect. 4 describes the announced data structure for paged memory.

2 Definitions

A binary relation $\mathcal{R}$ on a set S is defined as a set of pairs $\mathcal{R} \subseteq S \times S$; we write $a\mathcal{R}b$ if $(a, b) \in \mathcal{R}$. A relation $\mathcal{R}$ is said to be a *partial order* if it is:

- irreflexive, i.e. $(a, a) \notin \mathcal{R}$ for any a;
- transitive, i.e. $a\mathcal{R}b$ and $b\mathcal{R}c \Rightarrow a\mathcal{R}c$.

Order relations are usually denoted by the symbol $<$. Two elements a, b are said *incomparable* ($a \parallel b$) if neither $a < b$ nor $b < a$.

A partial order $\mathcal{L}$ is a *linear order* if for all pairs a, b we have $a\mathcal{L}b$ or $b\mathcal{L}a$; thus a linear order on a set S is the largest partial order relation that can be defined on S. An immediate consequence of the definition of linear order is that a linear order on n elements can be mapped into the usual "$<$" relation on the natural numbers $\{1, 2, \ldots, n\}$.

A classical measure of the complexity of a partial order is its *linear dimension*, first introduced in [7]. The linear dimension of a partial order $\mathcal{P}$ is defined as the minimum natural k such that there exist k linear orders whose intersection is $\mathcal{P}$.

More recently the concept of dimension has been generalized by using relations different from linear orders. The linear dimension is defined in terms if intersection of linear orders: if the base relation is required to be an interval order we obtain the concept of interval dimension, which is the minimum number of interval orders such that their intersection is the original partial order (see [10] for a definition of interval orders). An interesting generalization of the notion of linear dimension, interval dimension and several others is the *tube* dimension defined in [14].

An alternative measure of the complexity of a partial order is its *boolean dimension* [11], which is based on a more general composition of linear orders.

It is immediate to see that a partial order with linear dimension k can always be mapped into dominance in $\mathbb{R}^k$.

Given two elements a, b in a partially ordered set X we say $c \in X$ is their *least upper bound* ($\mathrm{LUB}(a, b)$) if:

- $a < c$ and $b < c$ and
- for each $d \neq c$ if $a < d$ and $b < d$ then $c < d$.

Analogously, given two elements a, b in a partially ordered set X we say $c \in X$ is their *greatest lower bound* ($\mathrm{GLB}(a, b)$) if:

- $c < a$ and $c < b$ and
- for each $d \neq c$ if $d < a$ and $d < b$ then $d < c$.

A partially ordered set is a *partial lattice* if any pair of elements has at most one least upper bound and at most one greatest lower bound.

If a partial lattice L contains a minimum, i.e. an element x such that $x < p$ for all $p \in L$, and a maximum, i.e. an element y such that $p < y$ for all $p \in L$, then L is called a *lattice*.

3 The Halfplane Search Problem

In what follows we show how to apply basic concepts in poset theory to the halfspace search problem. The approach we use for building the data structure described in Sect. 4 consists in explicitly storing the poset that represents the set of solutions, and navigating in it until the answer to the current query is found. A similar approach has been used in [12] for the stabbing problem on one-dimensional intervals.

Hence we first investigate the combinatorial structure of the set of possible answers to halfspace queries.

The *halfspace search* problem is classically stated as follows: given a set of n points $P \subseteq \mathbb{R}^d$, build a data structure that allows to efficiently determine which points lie in a closed query halfspace.

Halfspace search (usually called *halfplane search* in two dimensions) has been widely studied in two or more dimensions [3, 1, 15, 6, 4], both for points and for more complex objects like straight segments.

The halfspace search problem is particularly interesting since it represents a typical "non-isothetic" query on geometric data, and it is a basic primitive for many complex geometric operations. This "non-isothetic"-ness characteristic critically affects the design of a paged data structure, leading to solutions whose performances are very poor, with respect to either space or query time, while optimal solutions are available on the classical RAM model.

3.1 Halfplane Search and Circular Sequences

Edelsbrunner [8] provides a nice way of representing a set of points in the plane by a circular sequence. A *circular sequence* is a periodic sequence $\Pi_1, \Pi_2, \ldots, \Pi_m$ of permutations of integers $[1, 2, \ldots, n]$ such that

1. for each i Π_{i+1} is obtained from Π_i by reversing a number of disjoint substrings. We call this a *move*, and a *simple move* if the move swaps a single substring of length 2;
2. if a is swapped with b when passing from Π_i to Π_{i+1} and $j > i$ is minimal such that b is swapped with a when passing from permutation Π_j to Π_{j+1} then Π_j is the reverse of Π_i.

An immediate consequence of the above properties is that if permutation Π_i appears in a circular sequence $\mathcal{C}$ then also its reverse Π_i^R appears in $\mathcal{C}$; we call $\Pi_1, \Pi_2, \ldots, \Pi_1^R$ an *halfperiod* of $\mathcal{C}$. Property 2 above implies that any pair a, b swaps exactly once in each halfperiod; this plays a central role in many results concerning planar arrangements.

The link between circular sequences and halfplane search arises from the fact that for any set of points P in the plane a circular sequence $\mathcal{C}$ exists whose elements swap in the same way in which projections of points in P on a line l swap while continuously rotating l. In this case we say that $\mathcal{C}$ *realizes* P. Moreover if points in P are in general position (i.e. no three of them are collinear and no $a, b, c, d \in P$ exist such that line ab is parallel to line cd) then the associated circular sequence only contains simple moves.

In what follows we call k-set of $P \subset \mathbb{R}^2$ any subset $A \subseteq P$ such that $|A| = k$ and A is the answer to an halfplane query on P.

It is immediate to see that a k-set always is a prefix in some permutation of the realizing circular sequence, and a prefix in a permutation that realizes P always identifies a k-set of P. Hence any result on circular sequences can be transferred to k-sets, and we use this nice correspondence in order to prove most of the results in the next subsection.

3.2 Some Properties of k-sets

We investigate the structure of poset $(\mathcal{A}, \subseteq)$, where $\mathcal{A}$ is the family of all possible k-sets of a set $P, 1 \leq k \leq n$, determined by halfspaces containing point $x_1 = -\infty$, and $\subseteq$ is the set inclusion partial order relation.

A symmetry argument extends all our results, mutatis mutandis, to the poset defined by $\subseteq$ on the family of k-sets determined by halfspaces containing point $x_1 = +\infty$.

First of all we note that $|\mathcal{A}|$ is much smaller than 2^n.

Lemma 1. *The number of different solutions to halfspace queries in d dimensions is* $n^{\underline{d}} = n!/(n-d)!$.

Proof. Let $\pi_<$ be the subset of P contained in the interior of the lower (with respect to the first coordinate) halfspace defined by hyperplane π, and let $\pi_=$ be the subset lying exactly in hyperplane π, finally let $\pi_\leq = \pi_< \cup \pi_=$.

In order to prove the thesis, let us first disregard points lying in π, in the sense that we do not state whether points in $\pi_=$ belong to the answer set or not.

Suppose now a hyperplane π is given such that $|\pi_=| < d$: it is always possible to rotate π in π' until it hits one more point in $\pi_<$, giving $\pi'_= = \pi_= \cup \{p\}$ and $\pi'_\leq = \pi_\leq$. Thus any hyperplane defines the same answer set (disregarding points on the boundary) defined by a hyperplane containing exactly d points.

Since there are only $\binom{n}{d}$ hyperplanes containing exactly d points, and these points can be arranged in at most $d!$ ways producing $d!$ different solutions for hyperplanes passing just above or below each of them, the overall number of different solutions is not greater than $n^{\underline{d}}$. □

A different proof of Lemma 1 can be found in [8].

We study the characteristics of poset $(\mathcal{A}, \subseteq)$ in the case of n points in $\mathbb{R}^2$. In this case the problem is referred to as *halfplane search*, and the query is identified by a simple line.

Lower and upper bounds for the maximum number $e_k^{(2)}(n)$ of k-sets for any planar set of n points are provided in [9, 20]:

$$e_k^{(2)}(n) \in \Omega(n \log k), \text{ for } k \leq n/2,$$
$$e_k^{(2)}(n) \in O(n\sqrt{k}).$$

More results on $e_k^{(2)}(n)$ can be found in [8]. The upper bound has been recently improved to $O(n\sqrt{k}/\log^* k)$ in [17].

Restricting our attention to halfplanes containing point $x_1 = -\infty$ corresponds to taking into account only an halfperiod of the circular sequence. The remaining halfperiod of the circular sequence is exactly the symmetric. The basic consequence of this restriction is that any two elements in the circular sequence swap exactly once.

Theorem 2. $(\mathcal{A}, \subseteq)$ *is a lattice.*

Proof. Obviously P is the only maximum and $\emptyset$ is the only minimum in $\mathcal{A}$.

We now show that given $A, B \in \mathcal{A}$, with $A \parallel B$, there always exist a unique $\mathrm{LUB}(A, B)$ and a unique $\mathrm{GLB}(A, B)$ in $(\mathcal{A}, \subseteq)$. $\mathrm{LUB}(A, B)$ is the set determined by the common upper tangent [16] joining the convex hulls of A and B; it is immediate to see that it is the minimum k-set containing $A \cup B$. The unique

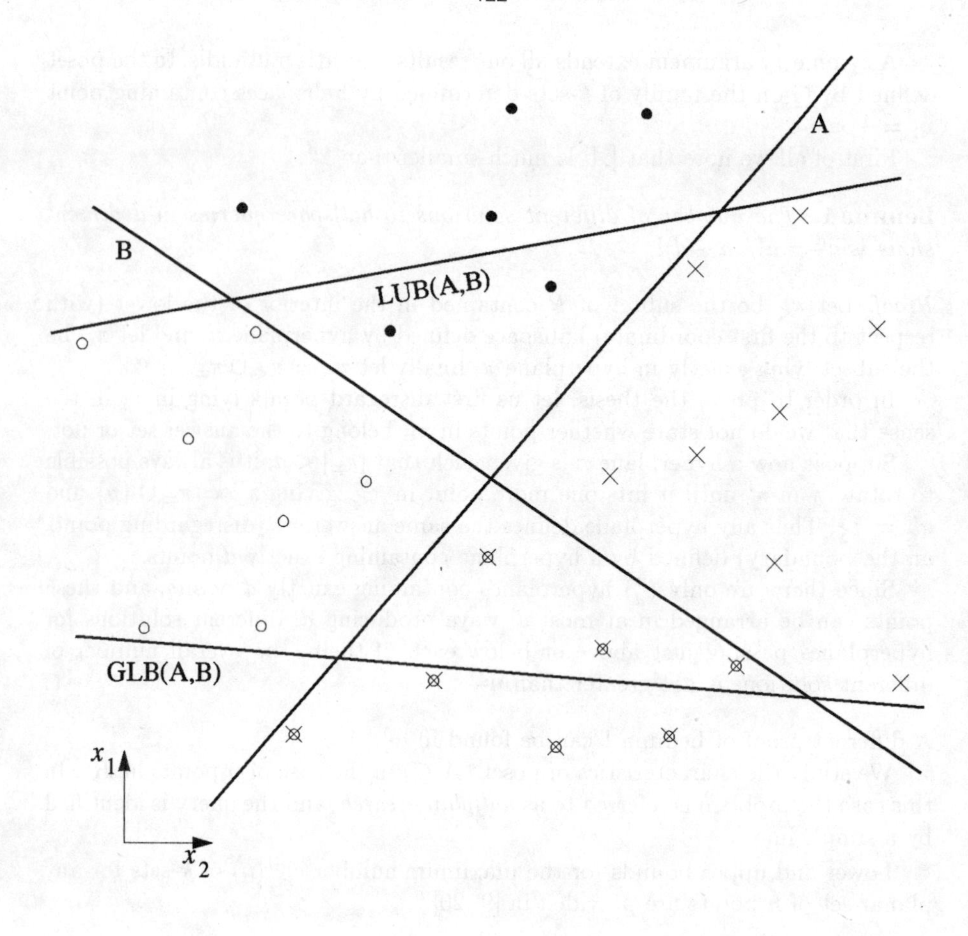

Fig. 1. Proof of Theorem 2

GLB(A, B) is the set determined by the common lower tangent [16] joining the convex hulls of $A - B$ and $B - A$. An example is shown in Fig. 1. □

The property stated in Theorem 2 does not hold in 3 dimensions, as can be seen looking at the set P consisting of two points p, q and $n-2$ points distributed on the surface of a small sphere centered on the middle point between p and q. Let A be the k-set containing p and some more point on the sphere, and let B be the k-set containing q and some more point on the sphere. It is simple to see that there are many LUB(A, B), and many GLB(A, B).

The following theorem states the main result in order to identify the complexity of $(\mathcal{A}, \subseteq)$.

Let us call $\mathcal{F}_A^k$ the family of all k-sets containing a set of points A.

Theorem 3. *For any fixed k, $1 \leq k \leq n$, it is possible to linearly order the* *family of k-sets on P such that given any i-set B, $i < k$, $\mathcal{F}_B^k$ is an interval in this ordering.*

Proof. Let $\mathcal{C}$ be a circular sequence realizing P, and let Π_x be a permutation in $\mathcal{C}$ having B as prefix.

For the sake of simplicity let us denote by $S_1, S_2, \ldots, S_m$ the (possibly coincident) k-sets defined by the first k elements respectively in $\Pi_1, \Pi_2, \ldots, \Pi_m$.

Obviously S_x contains B. We show that if S_y, $y > x$, does not contain B then all S_z, $z > y$ also do not contain B.

The proof proceeds by contradiction. Suppose S_y, $y > x$, does not contain B and let S_z, $z > y$, be the first k-set containing B, i.e. $S_{z-1} \not\supseteq B$. We color points in B *black* and all other points *white*. Hence there are a white point q and a black point p, respectively in the k-th and $k+1$-th position in Π_{z-1}, that are swapped between Π_{z-1} and Π_z. But p has already swapped with q between Π_x and Π_{z-1}, hence they cannot swap again in the same halfperiod.

The same argument applies to sets $S_{x-1}, S_{x-2}, \ldots$, thus proving the thesis.

□

The ordering in Theorem 3 is the same induced by the slopes of the halfplanes determining the k-sets, and it also is the order in which k-sets occur as prefixes in the circular sequence that realizes P.

Before stating the final result of this subsection, we need to show one more property.

Lemma 4. *For any couple of i-sets A, B of P, with A preceding B as prefix in the halfperiod of the circular sequence that realizes P, the first (resp. last) k-set in $\mathcal{F}_A^k$ precedes the first (resp. last) k-set in $\mathcal{F}_B^k$.*

Proof. Let us concentrate on the points that differentiate i-sets A and B: let a be the point in $A - B$ and let b be the point in $B - A$ (see Fig. 2). Points a and b swap between A and B, hence they cannot swap elsewhere in the halfperiod. Since the difference between the extension of $\mathcal{F}_A^k$ and $\mathcal{F}_B^k$ only depends upon the position of a and b the thesis trivially follows. □

The above properties ensure the following:

Theorem 5. $(\mathcal{A}, \subseteq)$ *has linear dimension 2.*

Proof. We show by induction on k how each k-set $S \in \mathcal{A}$ can be associated to a point $p(S) \in \mathbb{R}^2$ in such a way that $S \subset Q$ if and only if $p(S) \prec p(Q)$, where $\prec$ is the usual vector dominance relation in $\mathbb{R}^2$.

The basis of the proof consists in showing that for any fixed k the left-to-right order of representative points of k-sets corresponds to the order of k-sets defined in Theorem 3, and used in Lemma 4.

First of all 1-sets are mapped to points $(1, n), (2, n-1), \ldots, (c, n-c+1)$, in the order in which they appear as vertices of the lower hull [16] of P. Note that this is exactly the order defined by the slopes of lines that determine the 1-sets.

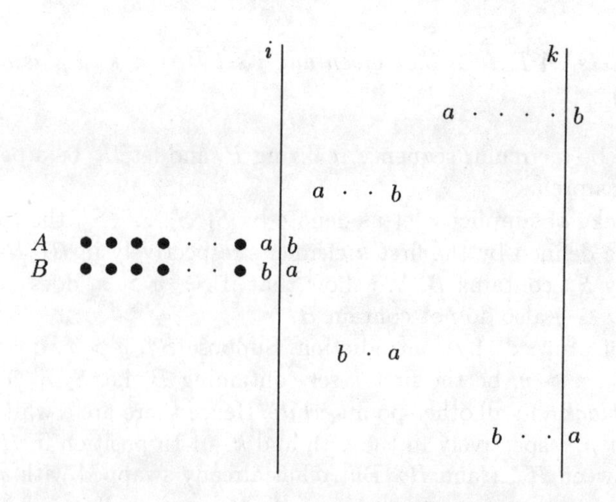

Fig. 2. Proof of Lemma 4

Assume now by induction that each k-set, $k \leq m+1$, has been assigned a point in $\mathbb{R}^2$ in such a way that $\subset$ is mapped to $\prec$, and representative points for $(m-1)$-sets appear in the order defined in Theorem 3: thanks to Lemma 4 it is possible to add representative points for m-sets according to the same ordering, as sketched in Fig. 3.

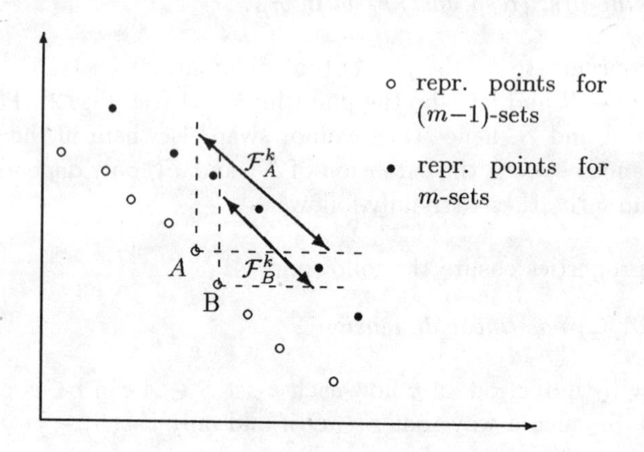

Fig. 3. Proof of Theorem 5

□

4 A Static Data Structure for Paged Memory

An optimal static solution to the halfplane reporting problem, on the RAM computational model, has been proposed in [4] based on the concept of convex layers [16]. It requires linear space and answers a query in time $O(\log n + k)$, where k is the cardinality of the reported set. Unfortunately this solution does not suggest any clustering technique that allows to obtain efficient structures for paged memory.

The computational model underlying the concept of paged memory consists of a memory organized in pages of size b. The space complexity is measured by the number of pages used, and the query time is measured by the number of pages accessed in order to answer a query.

An output sensitive solution on this model must access $\Theta(\frac{k}{b})$ pages in order to retrieve a solution consisting of k points.

We propose an output sensitive solution on this model whose qury time is $O(\log_b n \log\log n + \frac{k}{b})$, where k is the size of the answer and b is the number of points that fit in a page. The structure requires $O(\frac{n\sqrt{n}}{b})$ pages.

Our data structure is based on the idea of explicitly representing poset $(\mathcal{A}, \subseteq)$, and solving a query by navigating in the poset until the correct k-set is reached.

This brute force approach would lead to an $\Omega(n^2)$ space occupancy, since the number of different solutions can be $\Theta(n^2)$. We represent only k-sets whose cardinality is 2^i, $0 \leq i \leq \log_2 n$, and the navigation into $(\mathcal{A}, \subseteq)$ stops as soon as a set containing the solution is found. This set is at most twice as large as the answer, hence points that do not belong to the query halfplane can be discarded by means of a simple sequential scan, without increasing the asymptotic time complexity, thus recalling the *filtering search* technique introduced in [5].

On each layer, k-sets are organized according to the ordering defined in Theorem 3, and each k-set is labeled by the slope of the corresponding halfplane. This is needed in order to enter the layer in correspondence of the k-set that is the best candidate to contain the answer by means of logarithmic search.

The final trick for cutting down the space requirement comes from Theorem 3, and from the consideration that the union of t adjacent distinct k-sets has size not greater than $k + t$. This can be immediately seen by the fact, also showed in Fig. 2, that only one new point is added when passing from a k-set to the next one. So we group sets on layer 2^i in no more than $n\sqrt{2^i}/2^i = n/\sqrt{2^i}$ clusters, each representing 2^i adjacent 2^i-sets. The size of the union of sets in a cluster on layer 2^i is at most 2^{i+1}.

Clusters on layer 2^i are arranged in a B*-tree [2], whose leaves are labeled by the range of slopes corresponding to that cluster (note that these ranges define a partition of interval $[-\pi, \pi]$). Each cluster contains a pointer to a copy of its 2^{i+1} points.

We can thus state the following theorem.

Theorem 6. *It exists a data structure for the halfplane search on secondary storage that solves an halfplane query in* $O(\log k \log_b n + k/b)$ *worst case disk* *accesses and needs $O(n\sqrt{n}/b)$ pages, where k is the size of the output and b is the number of points that fit in a disk page.*

Proof. The total number of pages needed for representing all layers is

$$\sum_{i=0}^{\log_2 n} \frac{2^{i+1}}{b} \frac{n}{\sqrt{2^i}} \in O\left(\frac{n\sqrt{n}}{b}\right).$$

The answer to an halfplane query Q is computed by finding the cluster corresponding to the slope of Q on layers 2^i, for $i = 0, 1, \ldots$, and selecting points in the cluster that belong to Q. The process terminates as soon as a level is found such that the number of selected points does not increase with respect to the previous level, thus meaning that a cluster containing the answer to Q has already been found.

This process visits no more than $\lceil \log_2 k \rceil + 1$ layers, where k is the cardinality of the answer. Searching the right cluster and retrieving the corresponding points on layer i requires respectively $\log_b \frac{n}{\sqrt{2^i}}$ and $\frac{2^{i+1}}{b}$ disk accesses, thus giving

$$\sum_{i=0}^{\lceil \log_2 k \rceil + 1} \left(\log_b \frac{n}{\sqrt{2^i}} + \frac{2^{i+1}}{b} \right) \in O(\log k \log_b n + \frac{k}{b})$$

overall disk accesses. □

Note that the logarithmic overhead dominates the overall complexity only when small sets are reported, i.e. the answer has size $k = O(\log^2 n)$, thus the actual overhead is $O(\log_b n \log\log n)$.

Queries on halfplanes containing point $x_1 = +\infty$ can be answered by visiting a symmetric data structure.

This result is strongly meaningful with respect to any model of computation for secondary memory, and its relevance is due to the fact that the output sensitivity in the query time is in terms of pages.

5 Conclusions and Future Research

Poset theory seems to be a powerful tool to produce efficient data structures on paged memory for computational geometry problems, at the same time giving theoretical insight on the combinatorial structure of these problems.

An interesting open problem is to study the combinatorial structure of the answers to halfplane queries on complex objects (for example line segments or general convex domains). Some insight has been given by Santoro et al. [18, 19]; their results implicitly show that the structure of the poset underlying simple relations on non point-shaped objects can be arbitrarily complex, hence it is necessary to exploit more sophisticated concepts than the direct application of linear order dimension issues.

Another research direction consists in extending the set of primitives that can be applied on the data structure we propose.

References

1. Agarwal, P.K., Eppstein, D., Matoušek, J.: Dynamic half-space reporting, geometric optimization, and minimum spanning trees. Proc. 33rd Annu. IEEE Sympos. on Found. Comput. Sci. (1992) 80–89
2. Aho, A.V., Hopcroft, J.E., Ullman, J.D.: Data Structures and Algorithms. Addison-Wesley, Reading, MA, 1983
3. Brönnimann, H., Chazelle, B.,Pach, J.: How hard is half-space reporting. Discrete & Comput. Geom. **10** (1993) 143–155
4. Chazelle, B.,Guibas, L.J., Lee, D.T.: The power of geometric duality. BIT **25** (1985) 76–90
5. Chazelle, B.: Filtering search: a new approach to query-answering. SIAM J. on Comput. **15** (1986) 703–724
6. Clarkson, K.L., Shor, P.W.: Applications of random sampling in computational geometry, II. Discrete & Comput. Geom. **4** (1989) 387–421
7. Dushnik, B., Miller, E.W.: Partially ordered sets. American Journal of Mathematics **93** (1941) 600–610
8. Edelsbrunner, H.: Algorithms in Combinatorial Geometry. Springer-Verlag, Heidelberg, West Germany, 1987
9. Erdős, P., Lovász, L., Simmons, A., Straus, E.: Dissection graphs of planar point sets. In J.N. Srivastava editor, A Survey of Combinatorial Theory, North-Holland, Amsterdam, Netherlands, 1973, 139–154
10. Fishburn, P.C.: Interval Orders and Interval Graphs. John Wiley & Sons, New York, NY, 1985
11. Gambosi, G., Nesetril, J., Talamo, M.: On locally presented posets. Theoret. Comput. Sci. **68** (1990)
12. Giaccio, R., Talamo, M.: A general framework to deal with sets of intervals. In 8th National Conference on Logic Programming (GULP 93) Gizzeria Lido, Italy, 1993
13. Goodrich, M.T., Tsay, J., Vengroff, D.E., Vitter, J.S.: External-memory computational geometry. Proc. 34th Annu. IEEE Sympos. on Found. Comput. Sci. (1993) 714–723
14. Habib, M., Kelly, D., Möhring, R.H.: Comparability invariance of geometric notion of order dimension. Technical Report 320/1992, Technical University of Berlin, Berlin, Germany, 1992
15. Matoušek, J.: Reporting points in halfspaces. Comput. Geom. Theory & Appl. **2** (1992) 169–186
16. Preparata, F.P., Shamos, M.I.: Computational Geometry: an Introduction. Springer-Verlag, New York, NY, 1985
17. Pach, J., Steiger, W., Szemerédi, E.: An upper bound on the number of planar k-sets. Discrete & Comput. Geom. **7** (1992) 109–123
18. Santoro, N., Sidney, J.B., Sidney, S.J., Urrutia, J.: Geometric containment and partial orders. SIAM J. Disc. Math. **2** (1989) 245–254
19. Urrutia, J.: Partial orders and Euclidean geometry. In I. Rival editor, Algorithms and Order, Kluwer Academic Publishers, 1989, 387–434
20. Welzl, E.: More on k-sets of finite sets in the plane. Discrete & Comput. Geom. **1** (1986) 95–100

Triangle Graphs and Their Coloring

Yaw-Ling Lin

Department of Information Science
Providence University
200 Chung Chi Road, Sa-Lu, Taichung Shang, Taiwan
e-mail: yllin@host1.pu.edu.tw

Abstract. In this paper, we present results on two subclasses of trapezoid graphs, including simple trapezoid graphs and triangle graphs (also known as *PI* graph in [3]). Simple trapezoid graphs and triangle graphs are proper subclasses of trapezoid graphs [5, 3]. Here we show that simple trapezoid graphs and triangle graphs are also two distinct subclasses of trapezoid graphs.
Further, given the triangle representation and assuming that upper vertices of the triangles are listed in sorted order, we show that optimization problems on triangle graphs including finding the maximum independent set and minimum clique partition can be found in $O(n \log \alpha)$ time where α is the size of the largest independent set. The maximum clique and minimum vertex coloring in triangle graphs can be found in $O(n \log \chi)$ where χ is the chromatic number of the underlying triangle graph.

1 Introduction

In the literature of graph recognition algorithms, two of the most extensively covered graphs are interval graphs and permutation graphs. One way to generalize both interval graphs and permutation graphs is to consider two collections of intervals and define a one-to-one correspondence between them. Such graph is called the *trapezoid graph* [3, 4]. The fastest known algorithm for recognition of trapezoid graph is given by Ma and Spinrad in [10], where they show that trapezoid graphs recognition problem can be solved in $O(n^2)$ time.

On the other hand, Golumbic and Monma [8] introduced another class of graphs which generalizes the permutation graph and the interval graph. Recall that a graph $G = (V, E)$ is called a *tolerance graph* if for each vertex $v \in V$ there corresponds an interval I_v on a line and a positive real number t_v (the tolerance) such that $uv \in E$ iff $|I_u \cap I_v| \geq \min\{t_u, t_v\}$. Golumbic and Monma showed that, if all tolerances equal to a same constant, then we obtain exactly the class of all interval graphs. If the tolerances are $t_v = |I_v|$ for each vertex v, then we obtain the class of permutation graphs. Further, a tolerance graph is called *bounded* if $t_v \leq |I_v|$ for all $v \in V$. Bogart *et al.* [1] showed that bounded tolerance graphs are actually the *parallelogram* graphs; i.e., intersection graphs of parallelograms each of which has its horizontal lines on two parallel lines. Felsner [5] showed a trapezoid graph of 8 vertices that is not a tolerance graph (and thus not a bounded tolerance graph.)

The most interesting thing shown in the study of these subclasses of tolerance graphs is that, while they are all subclasses of trapezoid graphs, we still do not known how to efficiently recognize these graphs. This gives us a motivation of trying to find other properties of these subclasses of trapezoid graphs, which are still superclasses of permutation and interval graphs.

In this paper, the author presents results on two subclasses of trapezoid graphs including simple trapezoid graphs and triangle graphs. We call the intersection graph of rectangles and line segments whose two ends lie on two parallel lines the *simple trapezoid graph.* Since every rectangle and line segment are special cases of parallelograms, simple trapezoid graphs are also a proper subclass of trapezoid graphs. The intersection graph of triangles with top vertices on one line and bases on the other is called the *triangle graphs.* The triangle graphs were known by Corneil and Kamula as *PI* graphs. They showed that triangle graphs were a proper subclass of trapezoid graphs [3]. Surprisingly, such slight changes in the geometric structure render different intersection classes. In particular, we show that simple trapezoid graphs and triangle graphs are distinct from each other.

The second part of this paper concerns the algorithmic aspects of subclasses of trapezoid graphs. Dagan, Golumbic, and Pinter [4] showed that the channel routing problem is equivalent to the coloring problems on trapezoid graphs and they presented $O(n^2)$ algorithms for coloring and finding the maximum clique of trapezoid graphs. Felsner *et al.* [6] shows that these two problem can be optimally solved in $O(n \log n)$ time. In this paper, we give an algorithm that finds the maximum independent set and minimum clique partition in $O(n \log \alpha)$ time where α is the size of the largest independent set. Further, we show that finding the maximum clique and coloring triangle graphs can be done in $O(n \log \chi)$ time.

2 Simple Trapezoid Graphs

We start by considering the simplest generalization of interval graphs and permutation graphs. One way to generalize these two different graphs is to directly simulate their topological structures in the interval/permutation models.

Consider two parallel lines ℓ_1, ℓ_2 lie on the plane. We call a line segment whose endpoints one lies on ℓ_1 with the other lie on ℓ_2 as a *stick*, and call a rectangle whose two sides lie on ℓ_1, ℓ_2 a *block*. We then call the intersection graph of these sticks or blocks *simple trapezoid graphs.* Every simple trapezoid graph is a trapezoid graph since rectangles or line segments are just degenerate trapezoids. Simple trapezoid graph clearly generalize interval graphs and permutation graphs because interval graphs are the intersection graphs of blocks, and permutation graphs the intersection graph of sticks.

Note that C_4 is a permutation graph but not a interval graph, and also note that the graph F shown in Figure 1 is an interval graph but not a permutation graph. First observe that the induced subgraph of vertices $\{v_1, \ldots, v_6\}$ in G is a path P_6. Suppose that the graph F is a permutation graph, then the line segments corresponding to vertices $\{v_1, \ldots, v_6\}$ have a representation as shown

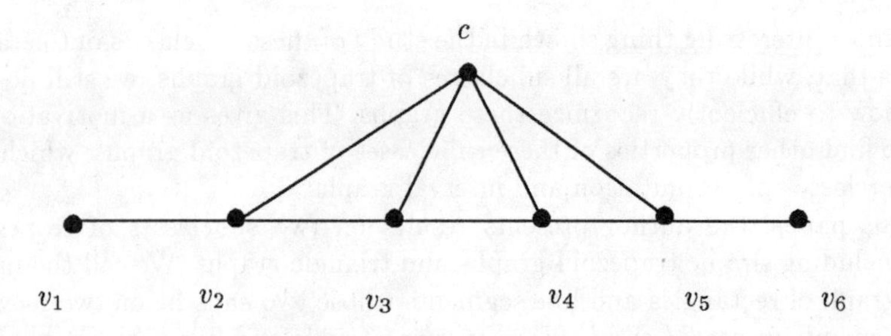

Fig. 1. The graph F is an interval graph but not a permutation graph.

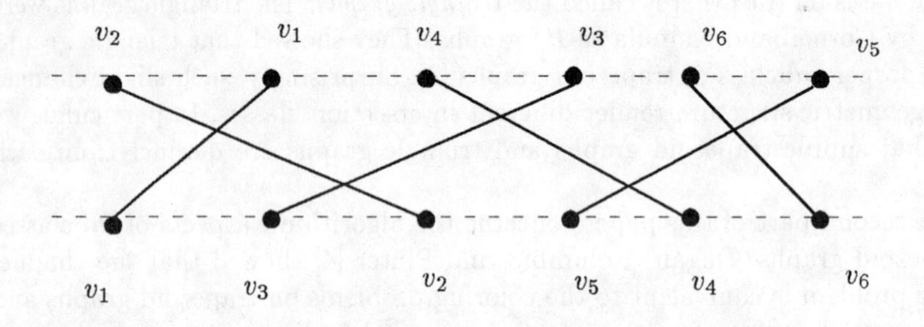

Fig. 2. The layout of vertices $v_1, \ldots, v_6$ of F in the permutation graph model.

in Figure 2 up to reflexive symmetry. However, then it is impossible to place a stick corresponding to the vertex c such that the stick will intersect exactly $\{v_2, v_3, v_4, v_5\}$. That is,

Lemma 1. *F is an interval graph but not a permutation graph.*

Note that the graph F is a *minimal* non-permutation graph, in the sense that deleting any vertex of F, the resulted graph is a permutation graph. Further, we can generalize the result of Lemma 1 to a family of graphs.

Again we use the notation $u \sim_G v$ to represent u and v are two adjacent vertices in G, i.e., $\{u, v\} \in E(G)$. When the underlying G is clear, we will drop the subscribe, and just write $u \sim v$. Further, given two subset of vertices A, B, we generalizes the notation in $A \sim B$ to mean that $a \sim b$ for all vertices $a \in A$ and $b \in B$. Let P_n be path of n vertices $\{v_1, \ldots, v_n\}$ such that $v_1 \sim \cdots \sim v_n$. We can then define F_n to be the graph including P_n and one extra vertex c such that $c \sim \{v_2, \ldots, v_{n-1}\}$. Note that $F = F_6$. Using a similar argument, it can be verified that each graph of the family $\{F_{2k} : k \geq 3\}$ is an interval graph but not a permutation graph, and they are all minimal non-permutation graphs.

Sticks and blocks in the models of simple trapezoid graphs do not have the same flexibility as the ordinary trapezoids. In particular, consider a C_4, the chordless 4-cycle, in the simple trapezoid graph. It is clear that not every vertex of a C_4 can have a block representation because C_4 is not an interval graph. Further,

Lemma 2. *Given a simple trapezoid graph G, let vertices v_1, v_2, v_3, v_4 induce a C_4 in G, such that $v_1 \sim v_2 \sim v_3 \sim v_4 \sim v_1$. Then no two adjacent elements can both have a block representation in the underlying simple trapezoid model.*

Proof. Suppose otherwise. Without loss of generality, assume that both v_1 and v_2 have block representations in the underlying model. None of the objects of v_1 or v_2 can contain each other in the model since $N(v_1) \not\subset N(v_2)$ and $N(v_2) \not\subset N(v_1)$. Without loss of generality, we can assume that some portion of the block corresponding to v_1 lies to the left of the block of v_2. Thus v_3 has to touch v_1 on the left, and now it is impossible for v_4 to touch both v_1 and v_3 without contact with the block v_2. □

The *Join* of two graphs G and H, denoted as Join(G, H), is the graph with $V(G) \cup V(H)$ as its vertex set, and $E(G) \cup E(H) \cup \{uv : u \in V(G), v \in H(G)\}$ its edges set. Now we are ready to show that not every triangle graph is a simple trapezoid graph:

Theorem 3. *Let F denote the graph illustrated in Figure 1.* Join(F, F) *is a triangle graph but not a simple trapezoid graph.*

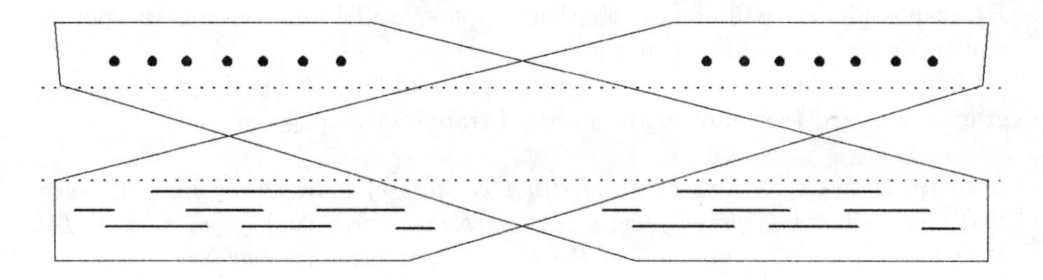

Fig. 3. A triangle model of the graph Join(F, F).

Proof. It is not hard to show that Join(F, F) is a triangle graph. Since F is an interval graph, we can take the interval models of F and duplicate these intervals into two copies. Put each copy in the lower line of the triangle graph model in the left and right corners, and associate the corresponding intervals in the lower left with the 6 user right points as well as the intervals in the lower right with 6 user left points to form 14 triangles. The triangle model of Join(F, F) is illustrated in Figure 3. It can be verified that the corresponding triangle intersection graph represented by these triangles is exactly the graph Join(F, F).

Partition the vertices of Join(F, F) into two groups $V = \{c, v_1, \ldots, v_6\}$ and $V' = \{c', v'_1, \ldots, v'_6\}$ such that $\langle V \rangle_{\text{Join}(F,F)} \cong \langle V' \rangle_{\text{Join}(F,F)} \cong F$, $v_1(v'_1) \sim$

$\cdots \sim v_6(v_6')$, and $c \sim \{v_2, \ldots, v_5\}(\{v_2', \ldots, v_5'\})$. Assume at first that the object corresponding to v_1 in the simple trapezoid model is a block. Since $v_1 \not\sim v_3$ in F, every two nonadjacent vertices of V' together with $\{v_1, v_3\}$ forms a C_4 in Join(F, F). Note that no vertex is adjacent to every other vertices in F. Thus, by Lemma 2, every vertex in V' must have a stick representation. However, since $\langle V' \rangle_{\text{Join}(F,F)} \cong F$ is not a permutation graph, we have to conclude that vertex v_1 can not be represented as a block in the underlying simple trapezoid model.

The argument that we used to conclude that v_1 has to be represented by a stick can be applied to every vertex of V (and V'.) This leads to a contradiction because F is not a permutation graphs. Thus we conclude that Join(F, F) is not a simple trapezoid graph. □

This result can be generalized to the family of F_n. That is, none of the graph in $\{\text{Join}(F_{2k}, F_{2k}) : k \geq 3\}$ is a simple trapezoid graph.

3 Triangle Graphs

A different approach to generalize interval graphs and permutation graphs is to consider the objects in each of the two parallel lines. We can consider the model in which the objects in one parallel line are intervals and the objects in the other line are points. We call such an intersection graph the *triangle graph* for the objects of this graph are triangles. Note that triangle graphs are also called *PI* graphs [3]. We will assume that the upper parallel line contains the points and the lower parallel line contains the intervals.

Before we show that the triangle graph is distinct from simple trapezoid graphs, we need the following property of trapezoid graphs:

Lemma 4. *Let $v_1, \ldots, v_6$ be six distinct vertices of a trapezoid graph G such that their induced subgraph form a $K_{3,3}$ such that $\{v_1, v_2, v_3\} \sim \{v_4, v_5, v_6\}$. Let S (T) be the middle trapezoid of the three trapezoids corresponding to vertices $\{v_1, v_2, v_3\}$ ($\{v_4, v_5, v_6\}$) in the trapezoid model of G. Then the upper (lower) interval of S is disjoint with the upper (lower) interval of T.*

Proof. Let t_i represent the trapezoid corresponding to the vertex v_i for $i \in [1..6]$ in the trapezoid model of G. Denote $t_i < t_j$ if t_i does not intersect with t_j and t_i is located to the left of t_j. Without loss of generality, we assume that $t_1 < t_2 < t_3$ and $t_4 < t_5 < t_6$; note that $t_2 = S$ and $t_5 = T$.

Assume that t_2 does intersect t_5 in the lower (upper) interval as illustrated by the diagram shown in Figure 4. Since t_2, t_5 intersect each other on the lower interval, the lower interval of t_4 lies on the left to the lower interval of t_3. By the same reason, the lower interval of t_6 lies on the right to the lower interval of t_1.

However, since $v_1 \sim v_6$ and $v_3 \sim v_4$, it implies that t_1 and t_6, also t_3 and t_4, intersect each other on the upper intervals, which is impossible for $t_1 < t_3$ and $t_4 < t_6$.

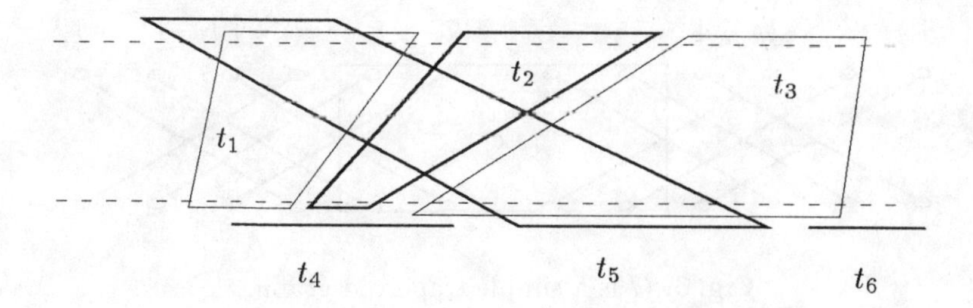

Fig. 4. A $K_{3,3}$ induced subgraph in trapezoid graph shall force a "X" shape of two middle trapezoids t_2 and t_5.

By symmetry, we reach the same contradiction if we assume t_2 and t_5 intersect in the upper intervals. That is, both the upper and lower intervals of T_1 and T_2 are disjoint to each other. In other words, they intersect each other by the "X" shape. □

Using this "X"-Shape Lemma (or the $K_{3,3}$ Lemma) as a gadget, we are able to design trapezoid graphs (actually better yet, simple trapezoid graphs) which do not have triangles representation. In particular,

Theorem 5. *The graph G shown in Figure 5 is a simple trapezoid graph, but not a triangle graph.*

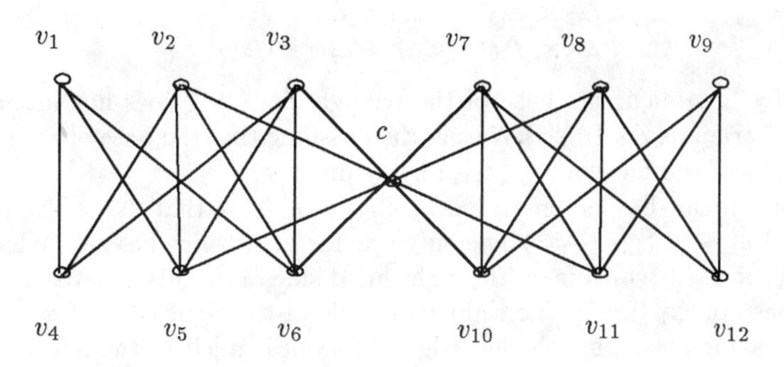

Fig. 5. A simple trapezoid graph which is not a triangle graph.

Proof. We first show that G is a simple trapezoid graph (thus a trapezoid graph) by presenting its model in Figure 6. Now we show that G is not a triangle graph.

Suppose that G is a triangle graph. Let Δ_i represent the triangle corresponding to the vertex v_i for $i \in [1..12]$, and let Δ_c represent the triangle of the vertex c.

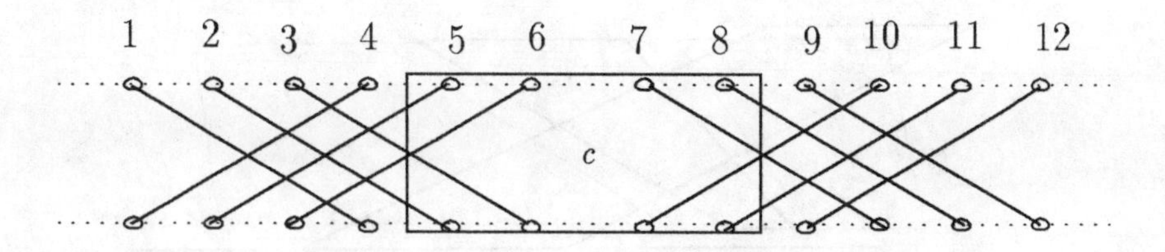

Fig. 6. G is a simple trapezoid graph.

By Lemma 4, the base of the middle triangle of $\{\Delta_1, \Delta_2, \Delta_3\}$ ($\{\Delta_7, \Delta_8, \Delta_9\}$) can not intersect with the base of the middle triangle of $\{\Delta_4, \Delta_5, \Delta_6\}$ ($\{\Delta_{10}, \Delta_{11}, \Delta_{12}\}$.) Furthermore, since the vertex c is adjacent to $\{v_2, v_3\}$ but not v_1, Δ_1 can not lie in the middle between Δ_2 and Δ_3. By the same reasoning, Δ_4 (Δ_9, Δ_{12}) does not lie between $\{\Delta_5, \Delta_6\}$ ($\{\Delta_7, \Delta_8\}$, $\{\Delta_{10}, \Delta_{11}\}$). That is, without loss of generality, we can assume that the triangle Δ_2 (and so Δ_5, Δ_8, Δ_{11}) lies between the two triangles $\{\Delta_1, \Delta_3\}$ ($\{\Delta_4, \Delta_6\}$, $\{\Delta_7, \Delta_9\}$, $\{\Delta_{10}, \Delta_{12}\}$.)

Further, note that vertices v_1 (or v_4), c, and v_9 (or v_{12}) form an independent set of size 3, and the vertex v_5 is adjacent to vertices v_1 and c but not to v_9 (or v_{12}). We conclude that Δ_c lies between the vertices v_1 (or v_4) and v_9 (or v_{12}.) Note that $v_1 \sim v_4$ and $v_9 \sim v_{12}$, so without loss of generality, we can assume that triangles Δ_1 and Δ_4 lies on the left hand side of Δ_c, while Δ_9 and Δ_{12} lies on the right hand side of Δ_c.

Again, let $\Delta_i < \Delta_j$ denote that the triangle Δ_i lies on the left hand side of the triangle Δ_j. From the previous analysis, we can now deduce the following relationships between these triangles:

$$\begin{matrix} \Delta_1 < \Delta_2 < \Delta_3 \\ \Delta_4 < \Delta_5 < \Delta_6 \end{matrix} < \begin{matrix} \Delta_7 < \Delta_8 < \Delta_9 \\ \Delta_{10} < \Delta_{11} < \Delta_{12} \end{matrix}$$

Recall that, by Lemma 4, the base of the triangle Δ_2 (Δ_8) does not intersect the base of the triangle Δ_5 (Δ_{11}.) We can then assume that the base of Δ_2 (Δ_8) lies on the right of the base of Δ_5 (Δ_{11}) by symmetry.

This situation can be demonstrated by Figure 7. Note that $\Delta_5 < \Delta_6$ (and $\Delta_7 < \Delta_8$) and $v_1 \sim v_6$ ($v_{12} \sim v_7$.) The only way for Δ_1 (which lies on the left of Δ_2) to intersect Δ_6 (which is on the right hand side of Δ_5) is to let the right edge of the base of Δ_1 lies on the right hand side of the right edge of the base of Δ_5. By the same reasoning, the left edge of Δ_{12} lies on left to the left edge of Δ_8. In short, since $\Delta_1 < \Delta_c < \Delta_{12}$, the base of the triangle Δ_c cannot intersect the base of triangles Δ_5 or Δ_8. But this is impossible, since that it implies that Δ_c can intersect with at most one of Δ_5 or Δ_8, but not both. However, in G, we have $v_5 \sim c \sim v_8$. Thus we reach the contradiction, and we conclude that G cannot be a triangle graph. □

Combining with Theorem 3, we have:

Corollary 6. *Triangle graphs and simple trapezoid graphs are two distinct proper subclasses of trapezoid graphs.*

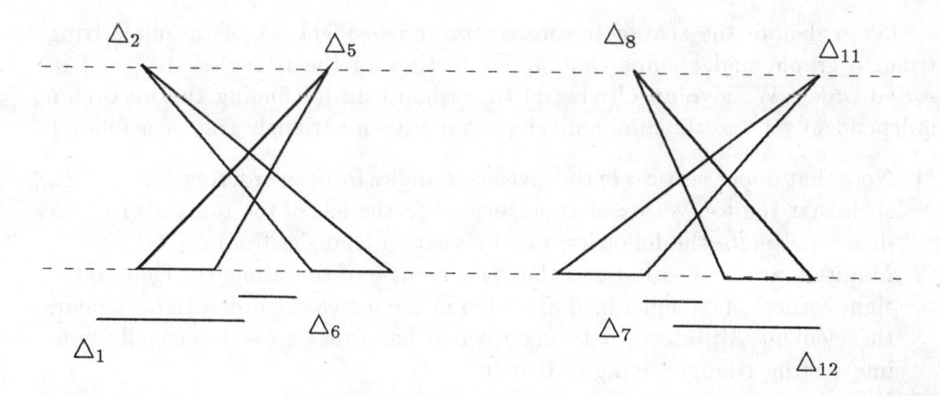

Fig. 7. G is not a triangle graph.

4 Optimization problems on Triangle graphs

Many optimization problems on triangle graphs can be solved efficiently since the class of triangle graphs is a subclass of trapezoid graphs, which is a subclass of incomparability graphs. Since comparability graphs are perfect [7], we can use the ellipsoid method to obtain the polynomial time algorithm for finding the maximum independent set, clique, chromatic number, and the clique partition number for triangle graphs.

In particular, given the transitive orientation of a comparability graph, we can find the maximum clique of a comparability graph in linear time [7]. Note that the maximum clique of a comparability graph corresponds to the longest chain of its corresponding transitive oriented graph, a partial order set, whose longest chain can be found in linear time by doing a topological sort [9]. Thus, given the triangle model, we can find the left-and-right relationships of triangles, which gives the partial ordering of the underlying triangles, and then we can use the linear time algorithm to find the longest chain in the poset. However, the algorithm would spend more than linear time even when the size of the edges set is linear ($O(n)$) because a naive approach in finding the ordering of these triangles still might need $O(n^2)$ comparisons for each pair of triangles.

4.1 Maximum Independent Set/Minimum Cliques Partition of Triangle graphs

One approach to find a subquadratic algorithm for finding the maximum independent set of trapezoid graphs involves sorting the upper coordinates of these triangles in $O(n \log n)$ time, and then finding the *immediate successor*, i.e., the *closest* non-intersecting triangle to the right, of each triangle in $O(\log n)$ time for a triangle. We say the triangle Δ_j is the *immediate successor* of a triangle Δ_i iff (1) Δ_j lies to the right of Δ_i, and, (2) among all triangles which lies on the right of Δ_i, the triangle Δ_j has a leftmost right corner.

Let α denote the size of the maximum independent set of the underlying triangle graph, and assume that upper vertices of the triangles are listed in sorted order. We give an $O(n \log \alpha)$ time algorithm for finding the maximum independent set and the minimum clique partition for triangle graphs as follows:

1. Note that upper vertices of the given n triangles form an ordering $\langle \triangle_1, \ldots, \triangle_n \rangle$ such that the top vertex of $\triangle_i$ is located to the left of the top vertex of $\triangle_j$ iff $i < j$. Iterate the following step by visiting triangles from $\triangle_1$ to $\triangle_n$:
2. Maintain an array A indexed the *right corners* of the triangles. That is, the right corners of triangles in A are listed in the increasing order. In particular, the element $A[i]$ stores a triangle which has exactly $i - 1$ mutually non-intersecting triangles lying to its left.
   Whenever a new triangle $\triangle_i$ is visited by the algorithm, we will check the position of the *left corner* of $\triangle_i$ by binary searching A. If $\triangle_i$ is the rightmost triangle so far, i.e., each triangle in the list lies on its left, we insert $\triangle_i$ into the array, which creates a new entry. Otherwise, $\triangle_i$ collides with some triangles in the list.
   Let $\triangle'$ be the leftmost triangle in the list collides with $\triangle_i$. Compare the right corners of $\triangle_i$ and $\triangle'$; the lesser one gets to stay, and the other one is thrown away. In particular, when triangles $\triangle_x$ and $\triangle_y$ collide with $\triangle_x$ discarded, we will record a *clique pointer* from the removed $\triangle_x$ to the accepted $\triangle_y$. Each triangle $\triangle_i$, if allowed to enter the triangle list, also keeps a pointer to its *immediate predecessor,* that is, the closest triangle on the list lying to the left of $\triangle_i$. Note that there are three different operations performed on the triangle list: (1) binary search the position of a point, (2) replace the value of an entry, and (3) insertion to the right of the list. Clearly an array suffices to do the job.
3. Note that triangles of A does not necessarily contain an independent set at every stage of the algorithm. Indeed, after all triangles are examined by Step 2, the algorithm looks for the last entry of the triangle list, and follows its predecessor link to report the independent set (longest chain) from right to left. To report the minimum clique partition, the algorithm scans each triangle and follows its clique pointer until it attains a sink triangle whose clique pointer is empty; triangles with empty clique are those on the triangle list. Note that triangles whose clique pointer reach the same sink can be viewed as an equivalent class; these equivalent classes together form the minimum clique partition.

So far we have completed the description of the algorithm. We claim that

Theorem 7. *This algorithm correctly reports a maximum independent set and a minimum clique partition of a triangle graph in $O(n \log \alpha)$ time given the triangle model where α is the size of the maximum independent set.*

Proof. For each triangle we need to spend $O(\log \alpha)$ for the binary search in Step 2. The final reporting of the maximum independent set in Step 3 takes $O(n)$ time, and the reporting of the minimum clique partition can also be done in $O(n)$ time by a depth-first-search algorithm since the underlying link structure is just a forest.

We now show that the algorithm correctly reports the maximum independent set. First, we claim that:

> A triangle $\triangle$ is ever accepted as the i-th entry of the triangle-list iff (1) the largest independent set with $\triangle$ as the rightmost triangle is of size i; (2) any triangle $\triangle'$ examined before $\triangle$ which also satisfies (1) has its right corner lies to the right of the right corner of $\triangle$.

When the first triangle enters the triangle-list, this claim is obviously true. By induction, we can assume that the claim holds for all the triangles whose upper vertices lie to the left of triangle $\triangle_j$. We will show that it is still true after $\triangle_j$ is examined by the algorithm. There are three cases:

Case 1. $\triangle_j$ is newly inserted at the i-th entry. $\triangle_j$ can not introduce an independent set of a larger size because by induction there are at most $i-1$ mutually nonintersecting triangles located to the left of $\triangle_j$. Condition (2) is also satisfied because no other triangle examined before $\triangle_j$ can introduce a size i independent set.

Case 2. $\triangle_j$ replaces the i-th entry $\triangle'$ in the list. By a same reasoning as the case 1, $\triangle_j$ does not introduce a larger sized independent set. Condition (1) is satisfied because its predecessor introduces a size $i-1$ independent set. Condition (2) is satisfied because, otherwise, $\triangle_j$ can not replace $\triangle'$.

Case 3. $\triangle_j$ collides with the i-th entry $\triangle'$, and $\triangle_j$ is discarded. $\triangle_j$ is not placed at the i-th entry because the Condition (2) for triangle $\triangle_j$ is false.

This completes the proof of our inductive argument. Thus, the last (rightmost) entry of the triangle list and its predecessor chain represent the largest independent triangle set by the claim.

We now prove that the algorithm reports a minimum clique partition. First we show that vertices of the each equivalent class defined by the clique pointer links indeed form a clique. Note that, when two triangle collide each other in Step 2, the removed triangle has its bottom interval contains the right corner of the accepted triangle. Thus two triangles pointing, by their clique pointers, to the same triangle share a common point. The only situation left to deal with is the case that two groups of triangles pointing to two different triangles in the same clique link. Note that points ever recorded at the same entry of the list always march from right to left, and the top vertices of the removed triangles march from left to right. It follows that the group sharing a common left point has its members' top vertices lie to the right of the other group which shares a common right point. Thus the triangles of these two groups mutually intersect each other. It follows that vertices of each equivalent class form a clique in the underlying triangle graph.

Now we show that it is a minimum partition. Since the underlying triangle graph contains an independent set of the same size, it needs at least as many cliques to cover all the vertices. Thus we conclude that the algorithm reports a minimum clique partition. □

Note that by this approach, it is not necessary to report the whole partial order set corresponding to the given model since the size of the underlying poset can be as large as $O(n^2)$.

Further, the scheme will also work nicely for permutation graphs when the lower intervals degenerate into points. The same algorithm can be used to find the maximum clique and vertex coloring in permutation graphs by reversing the ordering of points in the lower parallel line; the time complexity then becomes $O(n \log \chi)$ where χ is the size of the largest clique.

4.2 Maximum Clique/Minimum Vertex Coloring of Triangle graphs

We mentioned that the algorithm presented above can be used to find the maximum clique and minimum vertex coloring of *permutation* graphs in $O(n \log \chi)$ time. Can we also come up with an efficient algorithm for solving these two problems in triangle graphs? The answer is yes, as shown by the following algorithm:

1. Note that upper vertices of the given n triangles form an ordering $\langle \Delta_1, \ldots, \Delta_n \rangle$ such that the top vertex of Δ_i is located to the left of the top vertex of Δ_j iff $i < j$. Iterate the following step by visiting triangles from Δ_1 to Δ_n:
2. Maintain a priority queue, Q, of triangles using the x-coordinates of the *right corner* of each triangle as the keys. Note that the i-th largest entry of Q is the triangle Δ having the leftmost right corner such that if any incoming triangle Δ' (whose top vertex lies to the right of all triangles on the priority queue) has its left corner lying to the left of the right corner of Δ, Δ' will introduce a new clique of size at least $i + 1$.

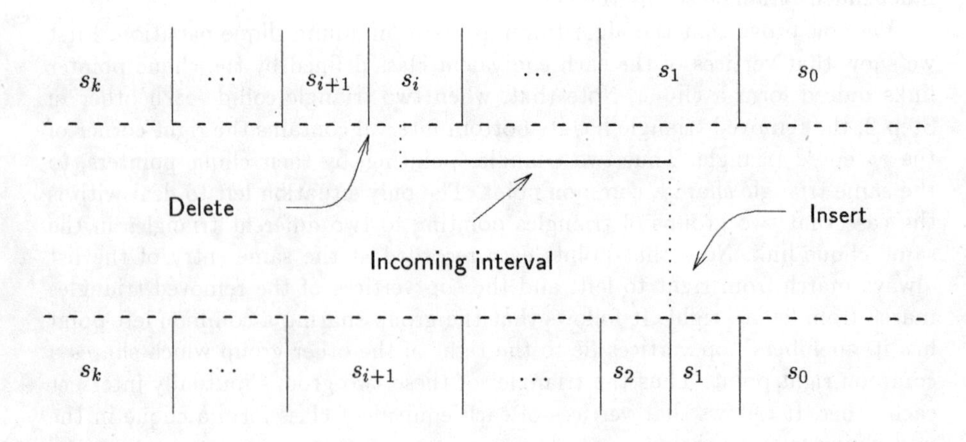

Fig. 8. The manipulation of the priority queue in finding the maximum clique of triangle graphs.

3. The right corners of triangles in Q define a partition of the horizontal line. Assume that a new triangle Δ is examined by the algorithm, let p_L (p_R) be the left (right) corner of Δ. Insert the triangle into Q using its p_R as the key, and delete the predecessor of p_L from Q if one exists. Note that the predecessor of p_L is the triangle lying to the left of p_L with the rightmost corner.
   Let the *predecessor pointer* of Δ be the removed triangle, Δ_L; Δ is given the same color as Δ_L's. If we cannot find Δ_L, Δ introduces a larger clique, and is assigned a new color; note that the size of the priority queue is increased by one. The step of this algorithm is illustrated by the diagram in Figure 8. Note that these right corners partition the horizontal lines into several sectors. Symbol s_i in Figure 8 thus denote the ith sector counting from the right.
4. Repeat Step 3 until all triangles have been examined. The size of the priority queue, $k = |Q|$, is the size of the maximum clique, and the chromatic number of the underlying triangle graph. Note that the each vertex/triangle is colored after the iteration terminates.
5. To report the maximum clique, find the leftmost entry in Q, and then visit all the triangles by the reverse order starting from Δ_n down to Δ_1. Let Δ_i be the currently visited triangle, and J_i the bottom interval of Δ_i. Let p be the leftmost point on the priority queue and K be the maximum clique initialized as an empty set. The possible relations between p and J_i are:

   Case 1: p is fully contained in J_i: Add i into the clique set, and remove the right endpoint of J_i from the priority queue.

   Case 2: p is the right endpoint of J_i: Add i into the clique set, remove p from the priority queue, and then replace p by p's next right hand side point in the priority queue. The iteration terminates if the priority queue becomes empty.

   Case 3: p is located to the left of J_i: Let Δ_j be the predecessor of Δ_i by following the predecessor link. Remove the right endpoint of J_i from the priority queue, and insert the right corner of Δ_j into the priority queue.

   Case 4: p is located to the right of J_i: Ignore J_i, and continue the iteration.

   The clique set K contains the indices of the maximum clique after the iteration.

Theorem 8. *This algorithm correctly reports the maximum clique and the minimum vertex coloring of a triangle graph in $O(n \log \chi)$ time given the triangle models where χ is the size of the maximum clique.*

Proof. For each interval we will need to spend at most $O(\log \chi)$ time for the insertion and the deletion of the priority queue, which can be implemented using a balanced binary tree, e.g., the R-B tree [2]. The colors the triangles is set in Step 3 in $O(n)$ time, and the last step of the algorithm for reporting the clique takes $O(n)$ iterations with each iteration spending at most $O(\log \chi)$ time for the priority queue manipulations; thus it also takes $O(n \log \chi)$ time.

The proof for the correctness of the algorithm is similar to the proof of Theorem 7. Here we first gives the proof that the algorithm gives a correct *size* of the maximum clique:

Suppose that, right before the algorithm examines a triangle with J as its bottom interval, there are already k points in the priority queue. These points partition the real line into $k+1$ sections. Denote these sections by $s_0, s_1, \ldots, s_k$ *from right to left*; see Figure 8. Assume that the following condition is true before the insertion of J:

> For any incoming triangle with bottom interval X, if the left vertex of X is located in the section s_i, then the largest clique including the incoming triangle has size $i+1$ for all $0 \leq i \leq k$ where k is the size of the priority queue.

Then we claim that after the manipulation of an interval J, it will remain valid. There are two cases:

Case 1: X intersects J. This means that the left endpoint of X lies to the left of the right endpoint of J. Therefore, if originally J introduces a clique of size q, now it will introduce a clique of size $q+1$ if these regions are also covered by J. Step 3 increases the indices of the sectors covered by J by one; thus the condition is still true after the insertion of interval J.

Case 2: X does not intersects J. This means that the left endpoint of X lies to the right of the right endpoint of J. Note the algorithm will not affect any sections on the left of interval J after the processing of interval J. Thus the condition remains true.

Thus in both case the condition remains true after the insertion of interval J. The condition is trivially true at the beginning of the algorithm. By induction, it follows that the algorithm reports the correct size of the maximum clique of the underlying triangle graphs.

The correctness of the vertex coloring follows from the fact that the algorithm reports a correct maximum clique. Suppose the algorithm reports a clique of size k. Then the algorithm assigns as many as k colors to these triangles. Note that the triangles in each predecessor chain form an independent set, which means the colors given by the algorithm is a valid coloring. Since the underlying triangle graph contains a clique of size k, it will at least need that many colors. Thus we conclude that the algorithm does report a minimum vertex coloring.

We now prove that K contains a valid maximum clique after the iteration of Step 5. First we show that elements of K indeed form a clique. The elements of K is obtained either from Case 2 or Case 3 of the Step 5. Before the leftmost point p is replaced in the Case 3, the triangles obtained from the Case 2 all share a common point, p; thus they form a clique. The only situation left to be dealt with is the relations between two groups of triangles originated from two different p's. Note that the point p is always marching from left to right, and the top vertices the triangles are visited *from right to left*. It follows that the group sharing a common left point has its members' top vertices lie to the right of the other group which shares a common right point. Thus the triangles of these two groups mutually intersect each other, which concludes that the elements of K induce a clique in the underlying triangle graph.

To prove that K is maximum, we need to show the size of K is k, the size of the maximum clique. Note that there are exactly k points on the priority queue when the iteration of Step 5 begins. Also note that whenever we add an element to K, the number of elements on the priority queue is decreased by one, and the algorithm does not decrease the size of the priority queue elsewhere. So it suffices to prove that every element on the priority queue is thrown away when the iteration finishes. However, it is certainly true since the iteration of Step 5 is essentially the reversed operation of Step 3. □

So far we have showed that finding the maximum independent set and minimum clique covering can be found in $O(n \log \alpha)$ time and finding maximum clique and chromatic number in $O(n \log \chi)$ time. However, we still seek an efficient algorithm for recognizing/inverting triangle graphs.

5 Acknowledgments

The author would especially like to thank Steven S. Skiena for the discussions about the properties of the subclasses of parallelogram graphs, and for the discussions of the optimization problems of triangle graphs. He gave the name of simple trapezoid graphs. The author thank Jeremy P. Spinrad and Tze-Heng Ma for the valuable discussions about the recognition of two dimensional interval orders and trapezoid graphs. The author would also thank Kenneth P. Bogart and Stefan Felsner for the discussions of the properties of bounded tolerance graphs and two dimensional interval order.

References

1. K.P. Bogart, G. Isaak, L. Langley, and P.C. Fishburn. Proper and unit tolerance graphs. Technical Report 91-74, DIMACS, November 1991.
2. T. Corman, C. Leiserson, and R. Rivest. *Introduction to Algorithms.* MIT Press, 1990.
3. D.G. Corneil and P.A. Kamula. Extensions of permutation and interval graphs. In *Proc. 18th Southeastern Conference on Combinatorics, Graph theory and Computing*, pages 267–276, 1987.
4. I. Dagan, M.C. Golumbic, and R.Y. Pinter. Trapezoid graphs and their coloring. *Discr. Applied Math.*, 21:35–46, 1988.
5. S. Felsner. Tolerance graphs and orders. In E.W. Mayr, editor, *Proc. 18th Internat. Workshop Graph-Theoret. Concepts Comput. Sci WG 92*, Lecture Notes in Computer Science 657, pages 17–26. Springer-Verlag, 1992.
6. S. Felsner, R. Müller, and L. Wernisch. Optimal algorithms for trapezoid graphs. Technical Report 368, Technische Universität Berlin, 1993.
7. M.C. Golumbic. *Algorithmic Graph Theory and Perfect Graphs.* Academic Press, New York, 1980.

8. M.C. Golumbic and C.L. Monma. A generalization of interval graphs with tolerances. *Congr. Numer.*, 35:321–331, 1982.
9. D. E. Knuth. *Fundamental Algorithms*, volume 1 of *The Art of Computer Programming.* Addison-Wesley, Reading, MA, 1968.
10. T.-H. Ma and J.P. Spinrad. An $O(n^2)$ time algorithm for the 2-chain cover problem and related problems. In *Proc. 2nd ACM-SIAM Symp. Discrete Algorithms*, pages 363–372, 1991.

Representation of an Order as Union of Interval Orders

Christian Capelle

LIRMM
UMR 9928 UNIVERSITE MONTPELLIER II/CNRS
161, Rue Ada
34392 Montpellier cedex 5 France
email: capelle@lirmm.fr

Abstract. We present an practical method to obtain a compact computer memory representation of orders and to compute pairwise comparisons efficiently. The principle of this method is to represent an order P as a union of interval orders P_i for which an optimal representation is already known (i.e. a *union representation of P* [Wes85]). For a directed acyclic graph $G = (X, U)$ representing an order $P = (X, <_P)$, the preprocessing time complexity is not better than the transitive closure computation cost. In the worst case, the size of the representation is the same that the size of the transitive closure. However, experimental tests give better results, and comparison with the compression technique of Agrawal & al. [ABJ89] is at the advantage of our method for dense orders.

1 Introduction

In computer science, there are many applications of orders. For example, in Objects Oriented Languages (OOL) the relation of subsumption is an order, as is the *is-a* (or *is-a-kind-of*) relation in Knowledge Representation Systems. In the future, in real applications, these hierarchies of objects will be very large. So, it will be very important to have efficient algorithmic techniques to represent them in memory and to compute object comparisons (see [HN] for a survey).

The ultimate goal would be to represent any order in a very small sized data structure and to have a very efficient algorithm to compare any pair of elements. But this problem seems to be very difficult.

In this paper, we expose an original method for representing any order as a union of interval orders (where an interval order is a kind of order such that an *optimal* representation is known) [Cap93]. This kind of representation technique, called *union representation*, defined at first by West in [Wes85], has not been studied very much neither on theoretical, nor in practical point of view, which was a motivation for our work. In fact, it is an original approach of order representation. We present theoretical performance evaluation of our method and experimental comparison with the compression technique of Agrawal & al. [ABJ89].

2 Definitions

A *finite partial order* (or more simply an *order*) $P = (X, <_P)$ is defined by a finite set X of *elements* or *vertices* and by a binary order relation $<_P$, i.e. an irreflexive, transitive and antisymmetric relation. If the relation is reflexive rather than irreflexive, it is denoted by $\leq_P$.

Two elements x and y of X are said to be *comparable* in P (denoted by $x \sim_P y$) iff $x <_P y$ or $y <_P x$. Otherwise, they are said to be *incomparable* (denoted by $x \|_P y$). The *cover relation* $\prec_P$ of P is defined by $x \prec_P y$ iff $x <_P y$ and there is no z in X such that $x <_P z <_P y$. If $x <_P y$, we say that xy is a *comparability* of P. The *number of comparabilities* of P is equal to the number of different pairs xy of X such that $x <_P y$.

We associate to P a unique directed acyclic graph $D(P) = (X, U)$ where $U = \{xy$ such that $x, y \in X$ and $x \prec_P y\}$. $D(P)$ is the *covering graph* or the *graph of the transitive reduction of* P. In this paper, by convention, figures will describe orders using their *covering graph*.

Let $G_1 = (X, U_1)$ and $G_2 = (X, U_2)$ denote two directed graphs. The graph $G_1 - G_2$ is defined by $G_1 - G_2 = (X, U)$ where $U = \{uv \in U_1$ such that $uv \notin U_2\}$. A directed graph $G = (X, U)$ is *transitive* iff for any pair of arcs xy and yz of U, xz belongs to U. The *transitive closure* of a directed graph $G = (X, U)$, denoted by $G^{tc} = (X, U^{tc})$, is the smallest (in the sense of the number of arcs) transitive graph including G. Its *transitive reduction*, denoted by $G_{tr} = (X, U_{tr})$, is the smallest (in the sense of the number of arcs) graph included in G, such that its transitive closure is G^{tc}.

We often do the misuse $P = G^{tc}$ where G_{tr} is the *covering graph* of P (i.e. $xy \in G^{tc}$ iff $x <_P y$). In the same way we can write $P = (X, U)$ where (X, U) is a transitive acyclic graph. Moreover, an order P can be interpreted by a directed acyclic graph $G = (X, U)$ such that $D(P) = G_{tr}$. We say that P is *represented* by the graph G.

$P' = (Y, <_{P'})$, the *suborder of* P induced by $Y \subseteq X$, is defined by ($x <_{P'} y$ iff $x, y \in Y$ and $x <_P y$), and denoted by $P_{|Y}$. $D(P')$ is a subgraph of $D(P)$. An order $P'' = (Z, <_{P''})$ is a *restriction* of P if $D(P'')$ is a partial graph of $D(P)$. The *union* of two orders $P = (X, <_P)$ and $P' = (Y, <_{P'})$ such that $X \cap Y = \emptyset$ is the order $P'' = (X \cup Y, <_{P''})$ such that $x <_{P''} y$ iff ($x, y \in X$ and $x <_P y$) or ($x, y \in Y$ and $x <_{P'} y$). The *intersection* of two orders $P = (X, <_P)$ and $P' = (X, <_{P'})$ is the order $P'' = (X, <_{P''})$ such that $x <_{P''} y$ iff $x <_P y$ and $x <_{P'} y$.

Throughout this paper, the following notations are used for a graph $G = (X, U)$: $ImSucc(x) = \{y \in X \ / \ xy \in U\}$, $TcSucc(x) = \{y \in X \ / \ xy \in U^{tc}\}$ and $TrSucc(x) = \{y \in X \ / \ xy \in U_{tr}\}$. $ImPred(x)$, $TcPred(x)$ and $TrPred(x)$ are defined similarly.

3 Orders Representation

A *representation* of a family of orders $\mathcal{F}$ is a memory representation for every order P belonging to $\mathcal{F}$, in a sense of a data structure or an abstract data type, such that, for any pair of elements x and y in P, we can answer this question: *"is x lower than y in P ?"*.

To compare two different representations of a family $\mathcal{F}$ of orders P, three complexity measures are generally considered:

- preprocessing cost (i.e. the cost, in time, to compute the representation),
- cost, in space, of the representation,
- cost, in time, of the predicate *"is x lower than y in P ?"*.

These complexity measures are in general computed considering the number of vertices and the number of edges in the graph representing the orders P of $\mathcal{F}$. They are not expressed in terms of the number of comparabilities of the order because, in actual applications, orders are often defined without all the comparabilities; in particular, most comparabilities obtained by transitivity, are not explicitly expressed.

In addition to these three complexity measures, others may be important to consider, depending on the application or on family of orders which is represented. For example, when lattice representations are studied, computation cost of the greatest lower bound (or smallest upper bound) of two elements is generally optimized rather than the pairwise comparison cost.

Finally, the possibility of updating incrementally the representation when the order is modified is a very important characteristic for real life applications.

What is a *good* order representation? Considering the family of orders such that their covering graph is an antichain, it is clear that this family can be represented in O(1) according to the three previous parameters. Such a representation can, of course, be qualified of *optimal.* However, this can be considered as a special case: Every family of orders with regular structure can be represented in constant space (description of the regular structure). So, our following definition of an *optimal* representation must be restricted to orders without regular structure.

An order representation of a family $\mathcal{F}$ of *non regular orders* is *optimal* if the three complexity measures presented above, expressed in terms of the size of the graph $G = (X, U)$ for all $P = G^{tc} \in \mathcal{F}$ are :

- preprocessing time cost: $O(|X| + |U|)$,
- the size of the representation: $O(|X|)$,
- O(1) time cost to compare two elements.

Universal Representations – Order Class Representations

Two categories of order representation methods can be distinguished :

The first contains representations intended to represent any order, but in a way not necessarily *optimal.* They are called *universal representations.* An universal representation tries to achieve a *good compromise*, depending on the type of application, between the three complexity measures to optimize. The method presented in this paper belongs to this category.

The second one, contains *optimal* representations for orders belonging to a *class* of orders. They are called *order class representations.*

Quick Overview of Order Representation Methods

In graph theory, order representations correspond to the representation of directed acyclic graphs (*dag* for short). Two abstract data types are well known and widely used to represent graphs in computer memory: matrices and adjacency lists. An order P can be represented by any graph whose transitive closure is P. These graphs belong to a range between $D(P)$ and $D(P)^{tc}$. According to the chosen graph and to the abstract data type, some of the complexity parameters can be improved, but not all. For example, a matrix representation of $D(P)^{tc}$ gives a representation with a constant time pairwise comparison cost while the size of the representation is quadratic even if the order contains few comparabilities. An adjacency list representation of $D(P)$ gives opposite results. Of course, more sophisticated techniques than the previous ones exist. Agrawal, Borgida and Jagadish developed a dag *transitive closure compression technique* [ABJ89] based on labelling spanning trees with numeric intervals, and then, adding in a bottom up manner, intervals corresponding to arcs not belonging to the spanning trees. This leads to a representation where x is lower than y if and only if the intervals set associated to x is included in the intervals set associated to y.
Bit vectors encoding techniques had been developed, in particular in Object Oriented Languages (or in Knowledge Representation Systems) where hierarchies induced by the relation of subsumption between objects are orders. These representation methods are often universal representation or sometimes restricted to lattices since in some applications hierarchies are lattices. Aït-Kaci & al. [AKBLN89] propose a labelling technique called *modulation*, embedding the order into a boolean lattice, leading to almost constant time lattice operations (but not in the worst case). Space complexity is quadratic. Caseau took and improved this idea in an incremental representation technique for the language Laure [Cas93]. It consists of transforming any hierarchy into a lattice which is then represented with a bit vector labelling technique called *compact hierarchical encoding*. This gives a constant time comparison cost while the representation size is quadratic. All these labelling methods can be denoted as *implicit* representations [KNR92] in the sense that the relation coding is totally included in the labels of vertices.

In distributed or parallel computing, the causality relation between events (for example, the event of sending a message on a processor precedes in the causality relation, the event of receiving that message by another processor) is an order. Lamport [Lam78], Fidge [Fid88] and Mattern [Mat89], have studied the representation of this order, in as accurate and efficient a way as possible. The causality order is maintained by exchanges of time stamps between processes. Fidge and Mattern propose an embedding of the causality order into the product of k chains (orders whose covering graph is a chain), where k is the number of processes, associating a vector of size k to each event. Every progress in the field of order representation may have applications in that field [Die92],[BF91].

In partial order theory, order representation methods have also been developed. Naturally, many classes have been studied. For some of them, optimal

representations are known (trivially for chains, antichains ..., but also for distributive lattices [Nou93]). If a class $\mathcal{F}$ is defined as a set of intersections (resp. unions) of orders belonging to a class for which a representation is known, then this give an *intersection representation* (resp. *union representation*) for $\mathcal{F}$. For example, dimension k partial orders (denoted by $dim(P) = k$) are orders defined by an intersection of k chains. This naturally leads to optimal representations for classes of bounded dimension orders. An important difficulty in developing representation methods based on dimension is that testing if $dim(P) \leq k$ is an NP-Complete problem for $k \geq 3$ [Yan82]. However, Ma & Spinrad [MS91] proposed an $O(n^2)$ recognition algorithm for two dimensional partial orders whose are given by a dag. Their technique allows the computation, in $O(n^2)$, of either the transitive closure or the transitive reduction of a two dimensional partial order.
In the same way, for any order P and any class $\mathcal{F}$, one can define the $\mathcal{F}$-dimension of P as the smallest number of orders belonging to $\mathcal{F}$ such that their intersection is P. For interval dimension two orders, the optimal space representation is computable in $O(n^2)$ from the transitive closure of the order [FHR91]. For some subclasses of interval dimension two orders, like bi-gap interval orders or proper bi-tolerance orders, optimal representations are computable in linear time [Bal94]. While intersection representations have been widely studied, union representations are less known. It is one of our motivations in studying a union representation method both on theoretical and practical points of view.
Madej and West [MW91] defined the *interval inclusion number*, $i(P)$, of an order P as the smallest t such that P has a mapping f in which each $f(x)$ is the union of at most t intervals, such that $\forall x, y \in P$, $x < y$ iff $f(x) \subset f(y)$. They give some theoretical results about this parameter: Trivially, $i(P) = 1$ if and only if $dim(P) \leq 2$. Interval orders are such that $i(P) = 2$ and there are orders of arbitrary high dimension such that their interval inclusion number equals 2. Testing $i(p) \leq k$ for any fixed $k \geq 2$ is NP-Complete. So, it is difficult to define representation methods based on this parameter. However, Agrawal & al. compression technique can be seen as a technique which constructs an approximation of the interval inclusion representation (a representation where the maximal number of intervals associated to an element is greater or equal to $i(P)$). This technique constructs an $i(P)$-sized representation (hence optimal) for orders such that their covering graph is a tree, however, the problem of having an universal $i(P)$-sized representation is NP-Hard.

4 Interval Restrictions Method

Theorem 1 presents a characterization of interval orders.

Theorem 1. *[Fis70, Möh89] Let $P = (X, <_P)$ be an order. The two following statements are equivalent:*

1. *P is an interval order.*
2. *P does not contain a suborder isomorphic to $2 \oplus 2$ (see Fig. 1)*

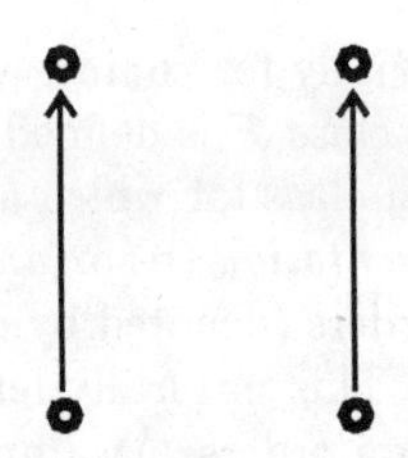

Fig. 1. the order $2 \oplus 2$

An *interval restriction* P' of an order P is an interval order which is a restriction of P. Interval orders constitute an order class for which *optimal* representation is known. Indeed, an interval order can be represented associating to each element x of X, an interval I_x of the real line in a way that :

$$x <_P y \Longleftrightarrow I_x \text{ is entirely to the left of } I_y$$

The size of this set of intervals has a linear complexity and is computable in linear time; moreover, the comparability of two elements can be computed in constant time.
The principle of our algorithm is the following :

Let $P = (X, <_P)$ be an order.
initialization : $i \longleftarrow 1$
$P_i \longleftarrow P$
Repeat
(1) computation of an interval restriction P'_i of P_i.
computation of an interval representation of P'_i.
(2) computation of the order $P_{i+1} \longleftarrow P_i - P'_i$
containing all the comparabilities of P_i not contained in P'_i.
$i \longleftarrow i + 1$
Until $(P_i = \emptyset)$

The algorithm returns a set of interval representations corresponding to the interval restrictions successively computed. To compare two elements of X, we have to search among the different representations, if there is one where the two elements are comparable.

Our algorithm constructs a *union representation* of P, i.e. a set of relations in a particularly nice class (here, interval orders) whose union is P. This set of relations is called a *union realizer* of P.

5 Computation of an Interval Restriction

To compute an interval restriction of any order P, we use the algorithm of P. Baldy and M. Morvan [BM93], recognizing if the transitive closure of a given directed acyclic graph is an interval order. We chose this algorithm because it is simple and efficient: its time and space complexity is $O(|X| + |U|)$ for a directed acyclic graph $G = (X, U)$. It is important to have an efficient algorithm because it is used several times in our method. Algorithm 1 describes the Baldy & Morvan algorithm.

Algorithm 1: IntervalOrder
Data: $G = (X, U)$ such that G^{tc} is an order.
Result: TRUE if P is an interval order, FALSE otherwise.
begin
  let Min be the set of minimal elements of G
  let S be the set of minimal elements of $G - Min$
  $interval\_order \longleftarrow TRUE$
  while $S \neq \emptyset$ **and** $interval\_order = TRUE$ **do**
    choose $x \in Min$ such that $|S \cap ImSucc(x)|$ is maximal
    if $|S \cap ImSucc(x)| \neq |S|$ **then** $interval\_order \longleftarrow FALSE$
    $X \leftarrow X - \{x\}$
    compute the new values of Min and S
  endwhile
  return $interval\_order$
end

The proof of this algorithm is a consequence of the following Theorem 2. The invariant of the While loop is $|S \cap ImSucc(x)| = |S| \Leftrightarrow interval\_order = TRUE$, just before x is taken out of X.

Theorem 2. *[BM93] Let $x \in Min$ such that $|S \cap ImSucc(x)|$ is maximal. Then P is an interval order if and only if $|S \cap ImSucc(x)| = |S|$ and $P - \{x\}$ is an interval order.*

We have modified the*IntervalOrder* algorithm to create an algorithm called IntervalRestriction (see Algorithm 2), which computes an interval restriction of any order P.

Proposition 3. *The algorithm IntervalRestriction(G) returns a graph G' such that G'^{tc} is an interval restriction of the order represented by the graph G.*

Proof. The invariant of the While1 loop remains True despite the modifications of the algorithm *IntervalOrder*. x is removed from X only if $|S \cap ImSucc(x)| = |S|$:
A chosen vertex x will be removed from X only when the condition of the invariant is verified. While it is not, the algorithm removes from the graph G' the incident arcs to some vertex y of S such that y is not immediate successor of x, then the new values of S and Min are computed, and finally the invariant condition is computed again. In fact, a vertex x is chosen only once and as many arcs as necessary are removed from G to insure the condition of the invariant; then x is removed from X. □

Algorithm 2: IntervalRestriction
Data: $G = (X, U)$ such that G^{tc} is an order.
Result: $G' = (X, V)$ with $V \subseteq U$ such that G'^{tc} is an interval order.
begin
  $G' \longleftarrow G \; \{V \longleftarrow U\}$
  let Min be the set of minimal elements of G'
  let S be the set of minimal elements of $G' - Min$
  while1 $S \neq \emptyset$ **do**
    choose $x \in$ Min such that $|S \cap ImSucc(x)|$ is maximal
    while2 $|S \cap ImSucc(x)| \neq |S|$ **do**
      for each $y \in S - ImSucc(x)$ **do**
        remove vertices zy from V
        compute the new values of Min and S
      endforeach
    endwhile2
    $X \leftarrow$X$-\{x\}$
    compute the new values of Min and S
  endwhile1
end

To analyze the complexity of *IntervalRestriction*, a detailed version of the algorithm, is presented below (see Algorithm 3). It comes from the detailed version of the Baldy & Morvan' algorithm. We use the following variables :

- Min is a boolean array and S a doubly linked list.
- $NbSuccInS(x)$, $NbPredInMin(x)$ and $NbPredNotInMin(x)$ are integer arrays which represent, respectively, the number of immediate successors in S of x, for x in Min, the number of immediate predecessors in Min and the number of immediate predecessors not in Min, for any vertex x not in Min.
- To allow the choice, in constant time, of one element in Min with the maximum of successors in S we maintain an array T of vertices lists such that $T[i]$ contains the list of vertices of Min with i immediate successors in S. Each non empty list $T[i]$ also contains the greatest index preceding i in T and the smallest index following i in T, such that the lists corresponding to these indexes are not empty. This array is managed in a way that a vertex is always inserted in $T[0]$, and it is moved from one list to one other step by step. Each one of these operations has a constant time cost.

Proposition 4 (Complexity of the algorithm). *The time complexity of the algorithm IntervalRestriction is $O(|X| + |U|)$ for any graph $G = (X, U)$ such that G^{tc} is an order.*

Proof. The graph is represented by adjacency lists such that the lists of successors and predecessors are crossed (i.e. for any arc xy, the element representing this arc belongs to the list of predecessors of y and to the list of successors of

Algorithm 3: IntervalRestriction (detailed version)
Data: $G = (X, U)$ such that G^{tc} is an order.
Result: $G' = (X, V)$ with $V \subseteq U$ such that G'^{tc} is an interval order.

```
begin
  V ⟵ U
  S ⟵ Ø
  for each x ∈ X do
      NbSuccInS(x) ⟵ 0
      NbPredInMin(x) ⟵ 0
      NbPredNotInMin(x) ⟵ |ImPred(x)|
  endforeach
  for each x ∈ X do
      if NbPredInMin(x) = 0 and NbPredNotInMin(x) = 0 then
          AppendToMin(x)
      endif
  endforeach
  while1 S ≠ Ø do
      remove x from the maximal index list of T
      while2 |S ∩ ImSucc(x)| ≠ |S| do
          for each y ∈ S − ImSucc(x) do
              { remove from V all the arcs zy : }
              for each z ∈ ImPred(y) ∩ Min do
                  remove the arc zy from V
                  NbSuccInS(z) ⟵ NbSuccInS(z) − 1
                  move z in T to the list preceding its present list
                     { i.e. If z ∈ T[k], move z from T[k] to T[k − 1]}
              endforeach
              NbPredInMin(y) ⟵ 0
              NbPredNotInMin(y) ⟵ 0
              S ⟵ S − {y}
              AppendToMin(y)
          endforeach
      endwhile2
      RemoveFromMin(x)
  endwhile1
end
```

x). So, when an arc obtained from the list of predecessors is removed, it is also automatically removed from the list of successors.

In *IntervalRestriction*, $S - ImSucc(x)$ is examined at each iteration of the While2 loop. But to examine only once the immediate successors of x the following technique is used: When the first iteration of the While2 loop is performed, a boolean array containing the immediate successors of x is computed; then S is reorganized in a way that the immediate successors of x are in the beginning

```
Algorithm 4: RemoveFromMin(x)
begin
    Min(x) ⟵ FALSE
    for each y ∈ ImSucc(x) do
        NbPredInMin(y) ⟵ NbPredNotInMin(y) − 1
        if NbPredInMin(y) = 0 and NbPredNotInMin(y) = 0 then
            S ⟵ S − {y}
            AppendToMin(y)
        endif
    endforeach
end
```

```
Algorithm 5: AppendToMin(x)
begin
    Min(x) ⟵ TRUE
    append x in the list T[0]
    for each y ∈ ImSucc(x) do
        NbPredNotInMin(y) ⟵ NbPredNotInMin(y) − 1
        if NbPredNotInMin(y) = 0 then
            S ⟵ S ∪ {y}
            for each z ∈ ImPred(x) do
                NbSuccInS(x) ⟵ NbSuccInS(x) + 1
                move x in T from its present list to the following
                    { i.e. If x ∈ T[k], move x from T[k] to T[k + 1]}
            endforeach
        endif
        NbPredInMin(x) ⟵ NbPredInMin(x) + 1
    endforeach
end
```

of S; a pointer to the first element of S which is not immediate successor of x is maintained. When the new value of the set S is computed, an element which is immediate successor of x is inserted at the beginning of S, and an element which is not immediate successor of x is inserted at the end of S. So, during the other iterations of While2 loop, the elements of $S - ImSucc(x)$ can be examined without examining $ImSucc(x)$. Hence, one element will be examined a constant number of times.

Moreover, for every vertex x of X, there is at most one call to $AppendToMin(x)$ and $RemoveFromMin(x)$. Finally, immediate successors of a vertex x are examined at most three times and immediate predecessors at most twice. The result follows.

□

Proposition 5. *Let $G = (X, U)$ be a transitive directed acyclic graph. The interval restriction G' of G computed by the algorithm IntervalRestriction is maximal in the meaning of inclusion.*

Proof. Let $G' = (X, V)$ be the result of the algorithm *IntervalRestriction(G)* for $G = (X, U)$ a transitive directed acyclic graph. Let us suppose that $G'' = (X, W)$ is an interval restriction of G such that $V \subset W$

i.e. $\forall xy \in V \; xy \in W$
and $\exists u, v \in X$ such that $uv \in U$, $uv \in W$, $uv \notin V$

Let us show that G'' must contain a subgraph isomorphic to $2 \oplus 2$ induced by four vertices of X including u and v, which refutes the assumption on G''.
uv has been removed from V by the algorithm *IntervalRestriction* when an element $x_0 \in Min_0$ was the chosen vertex at the beginning of the While1 loop number j.
Let us notice S_i et Min_i, the values of S and Min at the end of the While2 loop number i, inside the While1 loop where x_0 has been chosen.
Let $\{y_i \; i = 1..p\} = ImSucc(x_0) \cap S_0 \subset S_0$.
- Either $u \in Min_0$ and $v \in S_0$. Then there exists at least one $y_i = y_{i_0} \in ImSucc(x_0)$ such that $y_{i_0} \notin ImSucc(u)$ because $|ImSucc(x_0) \cap S_0|$ is maximal over Min_0. $y_{i_0} \notin TcSucc(u) - ImSucc(u)$ otherwise y_{i_0} would not be in S_0. uv is removed during the first While2 loop, therefore $v \notin ImSucc(x_0)$; moreover $v \notin TcSucc(x_0) - ImSucc(x_0)$ because otherwise, v would not be in S_0. So, the subgraph of G'' induced by x_0, y_{i_0}, u and v is isomorphic to $2 \oplus 2$, hence G'' is not an interval order.
- Or there exists in U a chain $s_0, s_1, \ldots, s_k$ $k > 1$ with $s_{k-1} = u$ and $s_k = v$, such that the arcs $s_{i-1}s_i$, $\forall i \; 1 \leq i \leq k$ have been removed during the iteration number i of the While2 loop inside the While1 loop where x_0 has been chosen. Let $s_{i-1} \in Min_i$ and $s_i \in S_i$. $u \in Min_k$ and $v \in S_k$, $v \in S_k - ImSucc(x_0)$, so $v \notin ImSucc(x_0)$; $v \notin TcSucc(x_0) - ImSucc(x_0)$ because otherwise v would not be member of S_k. $y_i \notin TcSucc(u) \; \forall i \; 1 \leq i \leq p$ because otherwise y_i would not be member of S_0. Therefore the subgraph of G'' induced by x_0, y_i, u and v is isomorphic to $2 \oplus 2$, hence G'' is not an interval order. □

6 The Order P as a Union of its Interval Restrictions P_i'

To compare two elements x and y of X efficiently (i.e. only examine once the intervals associated to x and y in each interval restriction), each comparability $x <_P y$ of P must be present in one interval restriction. But on the example of Fig. 2 (an order P represented by two interval restrictions P_1' and P_2' which union is P) we can remark that the comparability $2 <_P 6$ is not contained in P_1' or in P_2'. Indeed, the comparability $2 <_P 6$ is obtained from the union of P_1' and P_2' by transitivity between $2 <_{P_2'} 5$ and $5 <_{P_1'} 6$. To insure that pair-wise comparison can be computed without realizing transitive closure between the interval restrictions, we prefer to compute the transitive closure during the construction of the representation. So, our method begins by computing the transitive closure of the graph representing the order P. In the principle of our algorithm presented Sect. 4, the stage (2) consists of computing $P_{i+1} = P_i - P_i'$ where P_{i+1} contains *all the comparabilities of* P_i *not contained in* P_i'. We can

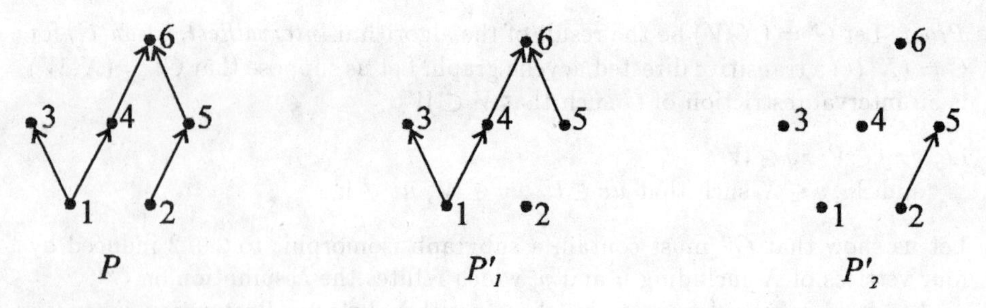

Fig. 2. An order $P = P'_1 \cup P'_2$ such that $P'_1 = IntervalRestriction(D(P))$ and $P'_2 = IntervalRestriction(D(P) - D(P'_1))$.

remark on Fig. 2, where orders are represented by non transitive graphs, that P'_2, computed by removing in the graph representing P the arcs of the graph representing P'_1, does not contain all the comparabilities of P not contained in P'_1. This is a consequence of the fact that all the comparabilities of P are not represented by an arc in his associated graph. Our algorithm computes a set of interval restrictions P'_i of P such that $P = \cup P'_i$. But this is not any union; indeed, it insures that each comparability of P is contained in one P'_i.

The transitive closure of the graph G representing the order P can be realized using the version of the algorithm of Goralcikova/Koubek [79] presented by M. Habib, M. Morvan and J.-X. Rampon in [HMR93, Mor91], whose time complexity, in the worst case, is $O(\mid X \mid + \sum_{y \in X} \sum_{x \in TrSucc(y)} \mid TcSucc(x) \mid)$.

An adjacency matrix representation of the transitive closure costs $O(n^2)$ in space and allows a constant time pairwise comparison. Even if in the worst case, our method seems no better than a simple matrix representation of the transitive closure (see Proposition 9), we will see that experimental results are quite better (see Sect. 9) which justifies the additional cost of our method with regard to transitive closure.

7 Interval Restriction Algorithm

The algorithm 6, *IRR* (Interval Restrictions Representation) is a detailed version of our representation method.

Proposition 6. *Let $G = (X, U)$ be a transitive directed acyclic graph representing an order $P = (X, <_P)$, then each comparability of P will be contained in one of the interval restrictions computed by the IRR algorithm.*

Proof. The *IRR* algorithm starts by computing the transitive closure of G, so each comparability of P is represented by an arc in G. At the beginning of the While loop number i, the graph $G - G'$ contains all the comparabilities of P (i.e. all the arcs of the initial graph G) not contained in the union $G'_1 \cup \ldots \cup G'_i$ of the already computed restrictions. So, at the end of the algorithm, each comparability of P will be in one interval restriction. □

Algorithm 6: IRR (*IntervalRestrictionRepresentation*)
Data: $G = (X, U)$ such that $G^{tc} = (X, U^{tc})$ is a partial order.
Result: For each element x of X, a list $intervals(x)$ of triplets $< i, lb, rb >$ where $[lb, rb]$ is the interval associated to x in the interval representation number i constructed by the algorithm *IntervalRestriction*. Each list is sorted according to the first element of the triplets.

begin
  $G \longleftarrow (X, U^{tc})$ { computation of the transitive closure of G }
  $i \longleftarrow 1$
  for each $x \in X$ **do**
    $intervals(x) \longleftarrow \emptyset$
  endforeach
  $G' \longleftarrow IntervalRestriction(G)$
    { first interval representation }
    { first assignment to *intervals*() lists }
  if $G = G'$ **then** $interval\_order \longleftarrow TRUE$ **else**
    $interval\_order \longleftarrow FALSE$
  endif
  while not($interval\_order$) **do**
    $G \longleftarrow G - G'$
    $i \longleftarrow i + 1$
    $G' \longleftarrow IntervalRestriction(G)$
      {*intervals*() lists are updated}
    if $G = G'$ **then** $interval\_order \longleftarrow TRUE$ **else**
      $interval\_order \longleftarrow FALSE$
    endif
  endwhile
end

Proposition 7. *The time complexity in the worst case, of the IRR algorithm, for any graph* $G = (X, U)$, *is:*

$$O(|X|^2 + |X||U^{tc}| + \sum_{y \in X} \sum_{x \in TrSucc(y)} |TcSucc(x)|)$$

Lemma 8. *Let* $G' = (X, V)$ *be the interval restriction computed by the algorithm IntervalRestriction on a transitive graph* $G = (X, U)$. *There exists at least one element* x *of* X, *such that all the arcs adjacent to* x *in* G, *belong to* G' *(i.e. there exists at least one element* x *of* X *such that all the comparabilities involving* x *in* G, *are in* G'*).*

Proof. Let x_0 be the first element chosen in Min by the IntervalRestriction algorithm. In the continuation of the algorithm, this choice is not reconsidered. The arcs whose tail is x_0 will not be removed from V in the While2 loop. After this loop, x_0 will be removed from X, and then, the arcs whose tail is x_0 will not be considered. As G is transitive, each comparability $x_0 <_G y$ is represented by an arc in G, so each of these comparabilities remains in G'. Moreover, x_0 is a minimal element in G, so G does not contain comparabilities of the form $y <_G x_0$. Hence, G' contains all the comparabilities involving x_0. □

Remark. The result of Lemma 8 is false if the graph G is not transitive.

Proof. (Proposition 7) The computation complexity of the transitive closure of G is $O(\mid X \mid + \sum_{y \in X} \sum_{x \in TrSucc(y)} \mid TcSucc(x) \mid)$.
Initialization of *intervals*() lists : $O(|X|)$.
Proposition 4 involves $O(|X| + |U^{tc}|)$ as the time complexity of *IntervalRestriction*.
Constant time computation of $G - G'$ because this computation is realized in *IntervalRestriction* without additional cost.
The test $G = G'$ can be realized in constant time.
So the time complexity of the While loop body is $O(|X| + |U^{tc}|)$. Lemma 8 implies that in the worst case, the number of iterations of the while loop is $|X| - 1$. The result follows. □

Remark. The result of Proposition 7 can be tempered by the fact that the worst case for the number of interval restrictions computed (i.e. $|X|$ interval restrictions), is a case where $|U^{tc}| \ll |X|^2$. So, an open question is to know if a more accurate complexity evaluation could give a better worst case result?

8 Memory Representation Size and Optimization

The size of the representation constructed by the *IRR* algorithm is equal to the number of elements of the *intervals*() lists. For each element x of X, *intervals*(x) contains as many triplets $< i, lb, rb >$, as there were calls to *IntervalRestriction*. Lemma 8 implies $O(|X|)$ calls to *IntervalRestriction*, hence Proposition 9 follows:

Proposition 9. *The size of the memory representation constructed by the IRR algorithm for a graph* $G = (X, U)$ *is* $O(|X|^2)$.

An optimization can be realized to reduce the representation size:
In the *IntervalRestriction* algorithm, for any element x of X, an interval is associated to x in each interval restriction, even for the interval restrictions where x is incomparable with any other element. It is easy to detect such elements and not to construct an interval for them. When the interval restriction number i is considered, if an element x does not have an interval number i, x is implicitly considered incomparable with any other element of X in this interval restriction. Figure 3 presents an interesting example of this optimization: The representation of the order $TWO_n = 2 \oplus 2 \oplus \ldots \oplus 2$ contains $2n$ intervals (2 intervals for each interval restriction P_i) rather than $2n^2$ without optimization. One open question is to know if Proposition 9 remains the same with the optimization. The worst example we know with the optimization gives an $O(n\sqrt{n})$ space representation. Perhaps, an improved evaluation of space complexity of our method might show a better worst case performance than $O(n^2)$.

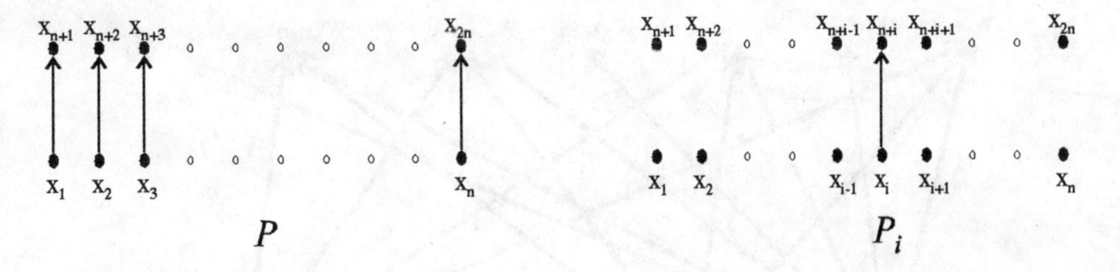

Fig. 3. The order $TWO_n = 2 \oplus 2 \oplus \ldots \oplus 2$ such that *IRR* must construct n interval restrictions $P_1, P_2, \ldots, P_n$.

9 Comparison with the Agrawal & al. Method

We tried to compare our method with other existing methods, in terms of performances. We have chosen the method of R. Agrawal, A. Borgida and H.V. Jagadish [ABJ89] because its practical performances are the best we have seen (theoretical performance of representation methods are often comparable in the worst case, so it is important to do practical comparison tests). In the following, we denote by *Labelling*, the algorithm of Agrawal & al.. We present first a theoretical comparison between the two methods, and then results obtained with an implementation.

If we consider the bipartite order of the Fig. 4, the size of the representation by the Labelling algorithm is $O(|X|^2)$, while the size of the representation constructed by the *IRR* algorithm is $O(|X|)$ (indeed, the example is an interval order). If we consider the order $P = TWO_n$, the size of the Labelling by the Agrawal & al. algorithm is $O(n)$, while the size of the representation constructed by the *IRR* algorithm is $O(n^2)$. For both of them, the preprocessing cost is bounded by the transitive closure cost. So, in general, none of these two methods is better than the other. We have implemented in C++ the interval restriction algorithm (*IRR*) and the transitive closure compression algorithm (*Labelling*). We have realized tests on random directed acyclic graphs G constructed in the following manner: We assign the number of vertices $n = |X|$ and the average number of successors of any vertex of G. The graph builder assigns randomly the successors of any vertex of X among the vertex in the range $[x+1, n]$ which guaranties that the resulting graph is acyclic.

The following array presents the results obtained on random graphs $G = (X, U)$ with varied values for $|X|$ and $|U|$. Each line gives average values obtained for ten random graphs (except five random graphs when $|X| = 1000$).

We note that both representations are smaller than the transitive closure graph, but greater than the transitive reduction graph.

For sparse orders (with few comparabilities), the representation constructed by the *Labelling* algorithm seems to have, in general, a smaller size than the representation constructed by *IRR*. More precisely the *Labelling* representation seems smaller than those of *IRR* in the ratio of one third. This can be explained by the fact that the *IRR* representation memorizes, with each interval, the num-

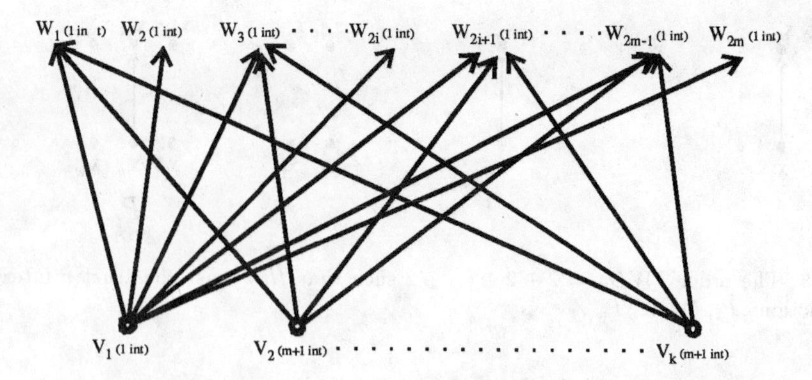

Fig. 4. An interval order P such that the Agrawal & al. representation is in $O(|X|^2)$ while the *IRR* algorithm representation is of size $O(|X|)$.

ber of the representation it belongs to (three integers rather than two). So this difference is a constant factor, not dependent on $|X|$.

$\|X\|$ (1)	$\|U\|$ (2)	$\|U^{tr}\|$ (3)	$\|U^{tc}\|$ (4)	*IRR* (5)	(6)	(7)	Labelling	(9)	(10)
50	221	83	740	341	2.3	8	199	2.0	11
50	398	78	970	254	1.7	6	197	2.0	13
50	594	76	1060	202	1.3	6	184	1.8	12
50	618	72	1044	215	1.4	5	167	1.7	12
100	448	200	2552	1017	3.4	17	563	2.8	18
100	888	193	3728	830	2.8	15	673	3.4	25
100	1237	186	4032	658	2.2	13	702	3.5	26
100	1599	176	4301	517	1.7	12	645	3.2	27
200	1884	474	13648	3170	5.3	29	2541	6.4	49
200	3689	426	16997	2094	3.5	26	2713	6.8	52
200	5092	371	17950	1298	2.2	23	2364	5.9	50
500	22733	1195	115463	8268	5.5	71	18762	18.8	138
500	37216	951	119408	3282	2.2	58	13374	13.4	130
500	31921	973	119014	3660	2.4	59	14648	14.6	122
1000	10527	3163	325296	70773	23.6	158	51732	25.9	208
1000	124463	2099	484825	12873	4.3	116	63530	31.8	266
1000	46944	2703	452267	35724	11.9	120	68414	34.2	238
1000	47382	2680	453886	34395	11.5	106	67538	33.8	230

Legend :

(1) $|X|$ =number of vertices in the graph G which is represented,

(2) $|U|$ =number of arcs in the graph G,

(3) $|U^{tr}|$ =number of arcs in the transitive reduction of G,

(4) $|U^{tc}|$ =number of arcs in the transitive closure of G,

(5) Size (the number of integers) of the representation constructed by *IRR*,

(6) Average number of intervals associated to any vertex x of X in the IRR algorithm representation,
(7) Total number of interval restrictions in the IRR algorithm representation,
(8) Size (the number of integers) of the representation constructed by the *Labelling* algorithm,
(9) Average number of intervals associated to any vertex x of X in the *Labelling* algorithm representation,
(10) Maximal number of intervals associated to a vertex of X in the *Labelling* algorithm representation.

As soon as the number of comparabilities goes to its maximum ($\sim 1/2|X|^2$), the representation computed by the IRR algorithm becomes smaller than the one computed by the *Labelling* algorithm (about twice as small). This can be explained by the more comparabilities the order contains, the lower is its probability of having many different suborders isomorphic to $2 \oplus 2$.

These good results are tempered by the fact that as soon as the number of comparabilities reaches $1/2|X|^2$, it can be more interesting to represent the complementary of the order rather than the order itself.

The complexity of the comparison of two elements x and y of P is proportional to the number of intervals associated with x and y in the representation. In this point of view, the IRR algorithm gives interesting results. In particular, the average number of intervals by vertex is clearly inferior to the one of the *Labelling* representation. So, our representation method gives very interesting result for representation size as well as for comparison cost. But its bad aspect is the impossibility to incrementally update the representation of P at a low cost when P is modified. For example, if we represent an interval order, we obtain only one interval restriction. We can modify the order by adding an element or removing a comparability in such a way that the new order is not an interval order. So, updating the representation becomes as difficult as a new computation of the representation.

We have seen that Agrawal & al. technique is an an approximation of the interval inclusion representation. In the same way one can associate to our method a parameter (called ***interval union*** parameter and denoted by $i_u(P)$), which associates to any order P the minimal number of interval orders such that their union is P. $i_u(P)$ is greater or equal to the greatest k such that TWO_k is a suborder of P. Trivially, $i_u(P) = 1$ iff P is an interval order, and there exist two dimensional partial orders for which interval union is arbitrarily high. An improved study of the ***interval union*** parameter seems to be an interesting question.

10 Conclusion

In this paper we develop an original approach of the order representation problem. Indeed, we use optimal representation techniques known for an order class, the interval orders, in the perspective of representing any order. We have shown that, although, in the worst case, performances are no better than transitive closure representation performances, experimental results are good. They are often comparable with Agrawal & al. method and, even sometimes better. The limit of our method comes from the difficulty to update incrementally the representation.

It would be interesting to study *union representation* techniques with other order classes (for example two dimensional partial orders). It would be nice to find a class which allows easy updating of the representation when deleting or adding comparabilities or vertices.

Acknowledgments

Special thanks to Michel Habib and Philippe Baldy for their many suggestions and our discussions in this work, and to the members of the "ALgorithmique COmbinatoire et LOgique" group for our discussions about the content of this paper.

References

[ABJ89] Rakesh Agrawal, Alex Borgida, and H.V. Jagadish. Efficient management of transitive relationships in large data bases, including is-a hierarchies. *ACM SIGMOD*, 1989.

[AKBLN89] Hassan Aït-Kaci, Robert Boyer, Patrick Lincoln, and Roger Nasr. Efficient implementation of lattice operations. *ACM Transactions on Programming Langages and Systems*, 11(1):115–146, january 1989.

[Bal94] P. Baldy. *Ensembles ordonnés: algorithmes, structures et applications aux systèmes distribués.* PhD thesis, Université Montpellier II, July 1994.

[BF91] Philippe Baldy and Christophe Fiorio. Estampillage des algorithmes distribués et ensembles ordonnés. Master's thesis, Université Montpellier II, 1991. Mémoire de DEA Informatique.

[BM93] Philippe Baldy and Michel Morvan. A linear time and space algorithm to recognize interval orders. *Discrete Applied Mathematics*, (46):173–178, 1993.

[Cap93] Christian Capelle. Représentation des ensembles ordonnés. Master's thesis, Université Montpellier II, June 1993. Mémoire de DEA Informatique.

[Cas93] Yves Caseau. Efficient handling of multiple inheritance hierarchies. In *OOPSLA'93*, pages 271–287, 1993.

[Die92] Claire Diehl. *Analyse de la relation de causalité dans les exécutions réparties.* PhD thesis, Université de Rennes I, 1992.

[FHR91] S. Felsner, M. Habib, and R.Möhring. On the interplay between interval dimension and ordinary dimension. In *Oberwohlfach Meeting on Ordered Sets*, Germany, 1991.

[Fid88] C.J. Fidge. Timestamps in message-passing systems that preserve the partial ordering, February 1988.

[Fis70] P. C. Fishburn. Intransitive indifference in preference theory: a survey. *Oper. Res.*, (18):207–228, 1970.

[HMR93] Michel Habib, Michel Morvan, and Jean-Xavier Rampon. On the calculation of transitive reduction-closure of orders. *Discrete Mathematics*, (111):289–303, 1993.

[HN] M. Habib and L. Nourine. Tree structure for distributive lattices and its applications. submitted to TCS.

[KNR92] S. Kannan, M. Naor, and S. Rudich. Imlicit representation of graphs. *SIAM J. Disc. Math.*, 5(4):596–603, November 1992.

[Lam78] L. Lamport. Time, clocks, and the ordering of events in distributed systems. *Communications of the ACM*, 7(21):558–565, July 1978.

[Mat89] F. Mattern. Virtual time and global states of distributed systems. In M. Cosnard and al., editors, *Parallel and Distributed Algorithms*, pages 215–226. Elsevier / North-Holland, 1989.

[Möh89] R. H. Möhring. Computationally tractable classes of ordered sets. In I. Rival, editor, *Algorithms and Orders*, pages 105–193. Kluwer Academic Publishers, 1989. volume : 255, series C: Mathematical and Physical Sciences.

[Mor91] Michel Morvan. *Algorithmes Linéaires et Invariants d'Ordres.* PhD thesis, Université Montpellier II, 1991.

[MS91] T.-H. Ma and J. Spinrad. Transitive closure for restricted classes of partial orders. *Order*, (8):175–183, 1991.

[MW91] T. Madej and D.B. West. The interval inclusion number of a partial ordered set. *Discrete Mathematics*, (88):259–277, 1991.

[Nou93] Lhouari Nourine. *Quelques Propriétés Algorithmiques des Treillis.* PhD thesis, Université Montpellier II, 1993.

[Wes85] Douglas B. West. Parameters of partial orders and graphs: packing, covering, and representation. In Ivan Rival, editor, *Graphs and Orders*, pages 267–350. NATO, D. Reidel publishing company, 1985.

[Yan82] M. Yannakakis. The comlexity of the partial order dimension problem. *SIAM J. Alg. Disc. Meth.*, 3:351–358, 1982.

Minimal Representation of Semiorders with Intervals of Same Length

Jutta Mitas*

Laboratoire d'Informatique de Robotique et de Microélectronique de Montpellier (LIRMM) 161 Rue Ada, F–34392 Montpellier Cedex 5, France
E-MAIL: mitas@lirmm.fr
and
TH Darmstadt, Fachbereich Mathematik, D–64289 Darmstadt, Germany

Abstract. Semiorders are special interval orders which allow a representation with intervals of same length. Using integer endpoints we present here such a representation with minimal interval length.
The algorithm obtained is linear in time and space. In addition, we give a characterization of the subclass of semiorders representable by intervals of length k by a set of C_{k+1} forbidden suborders where C_n is n-th Catalan number.

Keywords: Semiorder, interval order, minimal representation, forbidden suborders.

1 Introduction

Semiorders are an interesting subclass of interval orders, occuring for example in measurement theory. They can be used to model preference given by numerical values together with a constant tolerance limit for indifference, often called the 'just noticeable difference'.

Let us give the necessary definitions. In this paper, an ordered set $P = (X, <_P)$ consists of a finite ground set X and an irreflexive and transitive binary relation $<_P$ on X. P is called an *interval order* if it has an interval representation, i.e., a mapping that assigns to each $x \in X$ an interval $[l(x), r(x)]$ of the real line such that

$$x <_P y \iff r(x) < l(y) \; .$$

By Fishburn's theorem we have that an ordered set is an interval order if and only if it does not contain a suborder isomorphic to $2 + 2$ (see Figure 1, cf. [7] for more informations about interval orders).

The *predecessor set* of an element $x \in X$ is $Pred(x) := \{y \in X : y < x\}$. Similarly, the *successor set* of an element is defined by $Suc(x) := \{y \in X : y > x\}$.

* Supported by a postdoctoral fellowship of the Deutsche Forschungsgemeinschaft.

Fig. 1. The ordered sets $2+2$ and $3+1$.

Interval orders are characterized by the fact that the set of their predecessor sets as well as the set of their successor sets are linearly ordered by inclusion. A canonical interval representation with integers for interval orders which uses the minimal number of endpoints needed can be given by $x \longmapsto [l(x), r(x)]$ where

$$l(x) := |\{Pred(y) : Pred(y) \subset Pred(x)\}| \quad \text{and}$$
$$r(x) := |\{Suc(y) : Suc(y) \supset Suc(x)\}| \ .$$

A proof can be found for example in [9].

A *semiorder* $S = (X, <_S)$ is an interval order which allows an interval representation in which all intervals have the same size. A theorem by Scott and Suppes [13] tells us that an ordered set is a semiorder if and only if it does not contain a suborder isomorphic to $2+2$ or $3+1$ (see Figure 1). In addition, a semiorder has a, in some sense, canonical linear extension $L = x_1 x_2 \ldots x_n$ such that

$$Pred(x_1) \subseteq Pred(x_2) \subseteq \ldots \subseteq Pred(x_n) \qquad \text{and}$$
$$Suc(x_1) \supseteq Suc(x_2) \supseteq \ldots \supseteq Suc(x_n) \ .$$

The comparison between interval orders and semiorders with respect to their structural properties and the determination of order theoretical parameters reveals some interesting facts. A lot of problems have an easier access for semiorders as for interval orders.

The dimension of semiorders, for example, is never greater than 3 [11], whereas the dimension of interval orders is not bounded [12]. This yields on the one hand that the dimension of semiorders can be computed in polynmial time since it can be determined in polynomial time whether or not a given order has dimension 2. On the other hand, the complexity status of the dimension problem for interval orders is still open. The jump number problem is solvable in polynomial time for semiorders [1] but it is NP-complete for interval orders [9]. The 1/3–2/3–conjecture has been proven to be true for semiorders by Brightwell [5] but it is still open for interval orders. Up to now, nothing is known about the complexity status for counting the linear extensions of a semiorder or an interval order. Whereras in general, this problem is #P-complete [6].

We are searching for a common reason for the solvability of some problems for semiorders. But, on a first view, in all problems different properties of semiorders are used as a main constituent part for the solution. Nevertheless, representing a semiorder in levels according to the height of the elements reveals some important conditions for linear extensions. They cannot jump arbitrarily between the levels. More precisely, all elements of level k or below have to occur before any element of level $k+2$ or above. In the jump number problem this property helps to find the minimum number of jumps achievable by a linear extension. The 3 linear extensions which are needed in the dimension problem are proceeding level by level or wander at most between two consecutive levels before they move on to higher levels.

Recently, some work has been done about interval representations of semiorders with intervals of same length ([3],[10],[4]). We will call such representations *unit interval representations*. The canonical interval representation for interval orders does not produce in general intervals of the same length, even if it is applied to a semiorder. So it does not emphasize the typical properties of semiorders. Pirlot [10] determined a correspondence between the minimal interval length for a unit interval representation and the maximal number of edges in certain cycles of a valued graph.

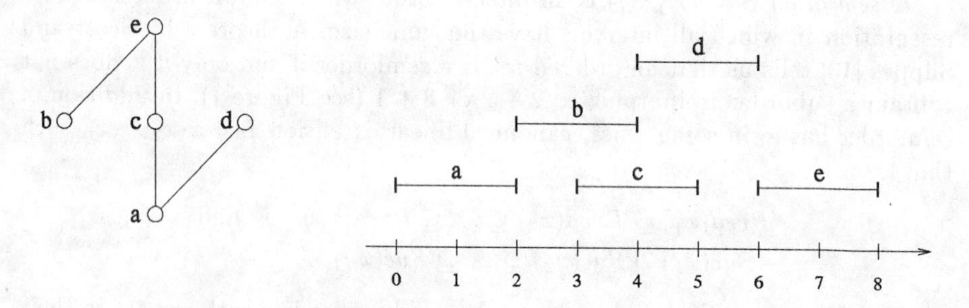

Fig. 2. A semiorder with a unit interval representation.

The purpose of this paper is to provide an unit interval representation for semiorders using integer endpoints. Moreover we will use the minimal length which is needed for the intervals under these assumptions. Thereby representing the semiorder in levels will again play an important role. For the representation we will introduce, in Section 3, an order T associated to the semiorder S and defined on the same elements as S. The introduction of T will be prepared in Section 2 making use of some of Pirlot's [10] ideas. The height of the elements in S and T will then suffice to present, in Section 4, the main result of the paper: A minimal unit interval representation of S. Our hope is that this somehow canonical representation for semiorders will help to understand this class better and maybe provide a method to attack yet unsolved problems.

In Section 5, we will show that finding this representation can be done in linear time. Finally, in Section 6, we will characterize by a set of forbidden suborders, the subclasses of semiorders that can be represented by intervals of length k, for some given natural number k. This characterization has been done also in the paper by Bogart and Stellpflug [4]. The example in Figure 2 shows a semiorder that is not representable by intervals of length 1 using integer endpoints, but by intervals of length 2.

2 Some Structural Properties of Semiorders

Let $S = (X, <_S)$ be a semiorder. A semiorder is called *reduced* if it does not contain elements x, y such that $Pred(x) = Pred(y)$ and $Suc(x) = Suc(y)$. Throughout this paper, we will consider only reduced semiorders since elements with the same predecessor set and the same successor set can get the same interval in the interval representation.

The height $h_P(x)$ of an element x of an ordered set P is the maximal size of a chain of elements below x. So, for example, the minimal elements have height 0. The height $h(P)$ of the order is the maximal size of a chain in P minus one.

Let L be the canonical linear extension of the semiorder S, i.e., $x \leq_L y$ if and only if $Pred(x) \subseteq Pred(y)$ and $Suc(x) \supseteq Suc(y)$. Certainly, we have

$$x \leq_L y \quad \Longrightarrow \quad h_S(x) \leq h_S(y) \ .$$

We will regard the semiorder represented in levels according to the height of their elements. In the sequel, the pairs of comparable elements of S which are as close as possible in L as well as pairs of incomparable elements of S which are as far as possible in L will play an important role (Pirlot [10] called these pairs 'noses' and 'hollows' because of their position in the incidence matrix). We define

$$(a,b) \in C :\Longleftrightarrow b = \min_L Suc(a) \qquad \text{and} \qquad a = \max_L Pred(b) \ ,$$
$$(b,a) \in I :\Longleftrightarrow a = \min_L X \setminus Pred(b) \qquad \text{and} \qquad b = \max_L X \setminus Suc(a) \ .$$

The following lemma shows that the arcs of C always climb one level, whereas the arcs of I either go one level down or stay on the same level.

Lemma 1. *Let a, b be two elements of the semiorder S. Then*
1. $(a,b) \in C \Longrightarrow h_S(b) = h_S(a) + 1$,
2. $(b,a) \in I \Longrightarrow h_S(b) - 1 \leq h_S(a) \leq h_S(b)$.

Proof. 1. Obviously, $(a,b) \in C$ implies $a <_S b$ and therefore $h_S(a) < h_S(b)$. On the other hand, there exists an element z with $z <_S b$ and $h_S(z) = h_S(b) - 1$. It follows $a \geq_L z$ and therefore $h_S(a) \geq h_S(z) = h_S(b) - 1$. Hence $h_S(a) = h_S(b) - 1$.

2. Because of the irreflexivity of S we have $b \not>_S b$ and therefore $a \leq_L b$. This yields $h_S(a) \leq h_S(b)$. Let now z be an element with $z <_S b$ and $h_S(z) = h_S(b) - 1$. It follows $a \geq_L z$, i.e., $h_S(a) \geq h_S(z)$. Hence $h_S(b) - 1 \leq h_S(a)$. □

3 The Order T

We now consider the transitive closure of the digraph on X obtained by joining the arcs of C with those arcs of I which go one level down. Formally defined

$$T := (C \cup I^*)^{tc}$$

where $I^* := I \setminus \{(x,y) | h_S(x) = h_S(y)\}$. See Figure 3 for an example. S is represented in levels and inside a level the linear extension L of S runs from the left side to the right side. Subsequently, we will show that T is an order. Later

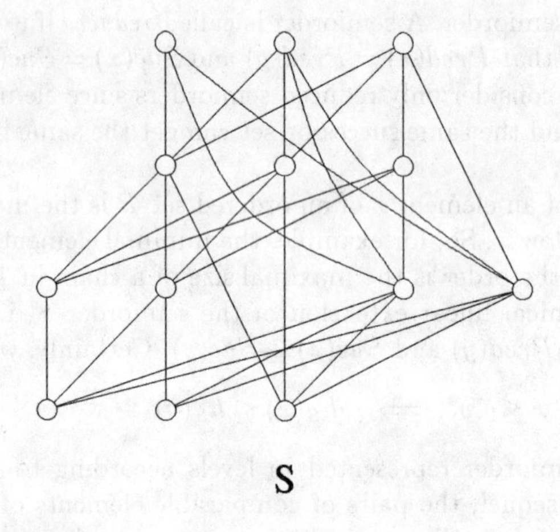

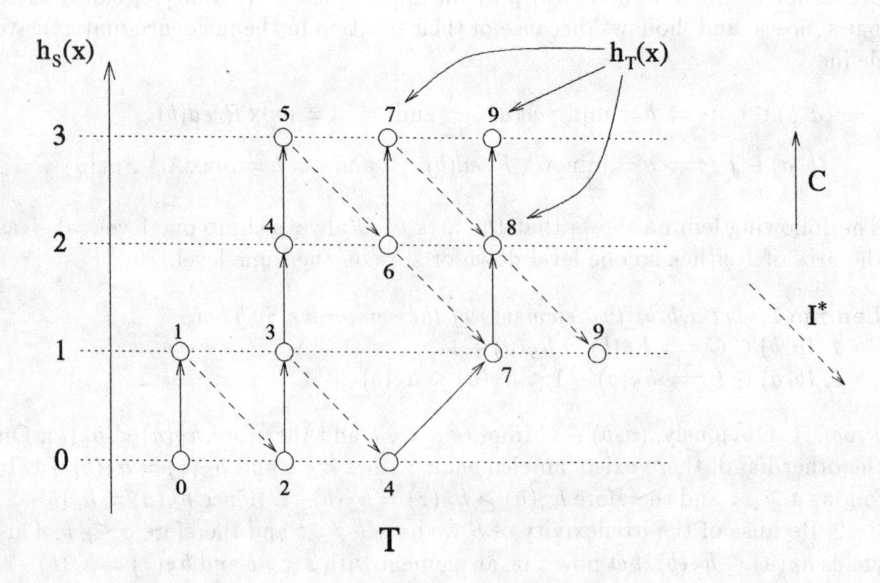

Fig. 3. A semiorder S and its representation by the order T.

on, we will see that the knowledge of the height of the elements in S and in T suffices to determine the intervals in an unit interval representation of S with intervals of minimal length.

Proposition 2. *T is an order.*

Proof. We have to prove that the digraph $C \cup I^*$ is acyclic. We will do this by showing that the existence of a directed path from a vertex x to a vertex y on the same level of S, implies that $x <_L y$.

Each arc of $C \cup I^*$ connects two consecutive levels of S. Therefore a path between two elements of the same level must have an even number of arcs. So, let for the beginning x, y be two elements of X with $h_S(x) = h_S(y)$ and there is a path from x to y of length 2. Then there exists z such that xI^*zCy or $xCzI^*y$. If xI^*zCy then $z \not<_S x$ but $z <_S y$. Therefore $x <_L y$. If, on the other hand, $xCzI^*y$ then $x <_S z$ but $y \not<_S z$. This implies $x <_L y$, too.

Now suppose that, for some i, it has been shown that for every path of length $2i$ which leads from a vertex x to a vertex y with $h_S(x) = h_S(y)$, it follows that $x <_L y$. Then consider a path of length $2(i+1)$ leading from a vertex a to a vertex b also satisfying $h_S(a) = h_S(b)$. If there exists an element y on the path from a to b with $h_S(y) = h_S(a) = h_S(b)$ then the two paths from a to y and from y to b have length not greater that $2i$ and therefore, by induction, $a <_L y <_L b$. If this path does not touch the level of a and b between them, then the vertex u right after a and the vertex v right before b on the path, satisfy either aI^*u and vCb or aCu and vI^*b. Hence $h_S(u) = h_S(v) = h_S(a) \pm 1$ and the path between u and v has length $2i$. Consequently $u <_L v$. If aI^*u and vCb then $u \not<_S a$ but $u <_S b$ since $u <_L v <_S b$. Therefore $a <_L b$. For the second case $a <_L b$ can be shown similarly.

Summarized, we obtain that for any $x \in S$ there can be no path in $I^* \cup C$ from x to x since this would imply $x <_L x$, a contradiction to the irreflexivity of S. Hence $I^* \cup C$ does not contain a cycle and T is an order. □

An important property of T is that there is always a path in T between two arbitrary elements which are on the same level of S. The path is leading from the smaller element in L to the greater element in L, as we will see in the following Proposition.

Proposition 3. *If $x <_L y$ and $h_S(x) = h_S(y)$ then $h_T(x) < h_T(y)$.*

Proof. Let x_i, x_{i+1} be two consecutive elements of L which are on the same level of S. The aim is to show $x_i <_T x_{i+1}$. This implies the claim.

Since we assume that S is reduced, x_i and x_{i+1} have different predecessor or different successor sets. Suppose, $\mathrm{Suc}(x_i) \neq \mathrm{Suc}(x_{i+1})$. Then let $a := \min_L \mathrm{Suc}(x_i) \setminus \mathrm{Suc}(x_{i+1})$ and $b := \max_L \mathrm{Suc}(x_i) \setminus \mathrm{Suc}(x_{i+1})$. It follows $(x_i, a) \in C$ and $(b, x_{i+1}) \in I$. Moreover, $b \in \mathrm{Suc}(x_i)$ and therefore $h_S(b) > h_S(x_i) = h_S(x_{i+1})$. This yields $(b, x_{i+1}) \in I^*$.

If $a = b$ then $x_iCaI^*x_{i+1}$, i.e., $x_i <_T x_{i+1}$. If $a \neq b$, note that all elements of $\mathrm{Suc}(x_i) \setminus \mathrm{Suc}(x_{i+1})$ have the same predecessor set and therefore different successor sets. Considering an arbitrary pair of consecutive elements y_i, y_{i+1} of

Suc(x_i)\ Suc(x_{i+1}), we get with the same arguments as above, elements u and v such that $(y_i, u) \in C$ and $(v, y_{i+1}) \in I^*$. This process can be continued until we reach the highest level of the order. Recall that each time we find two new vertices they will be one level higher as the elements to which they are connected. Since elements of the highest level cannot have different successor sets, they must be equal if they have the same predecessor set. Therefore a path from x_i to x_{i+1} in $C \cup I^*$ will be obtained. Hence $x_i <_T x_{i+1}$. If we suppose in the beginning that x_i and x_{i+1} have different predecessor sets then $x_i <_T x_{i+1}$ can be obtained analogously. □

4 Unit Interval Representation of Semiorders

Let S be a semiorder. For $i \in \{0, \ldots, h(S)\}$ define $a_i := \min_L\{x : h_S(x) = i\}$ and $z_i := \max_L\{x : h_S(x) = i\}$. Further let

$$d := \frac{1}{2} \max_{0 \le i \le h(S)} (h_T(z_i) - h_T(a_i)) .$$

We are now ready to state the main results.

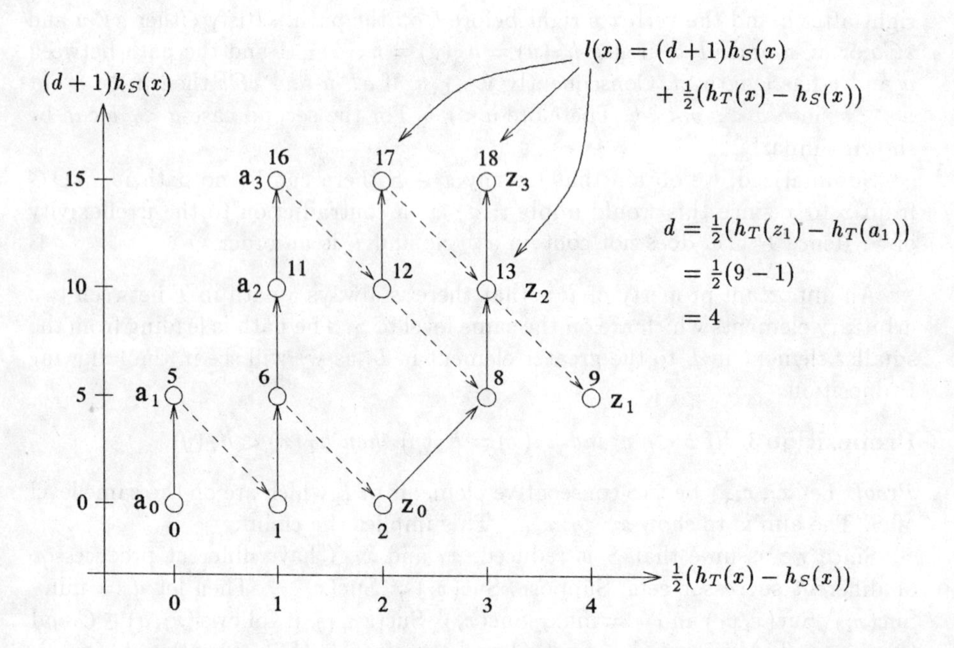

Fig. 4. Unit interval representation of S.

Theorem 4. *Let S be a semiorder. For $k \in \mathbb{N}$, $k \geq d$, the mapping $x \longmapsto [l(x), l(x)+k]$ where*

$$l(x) := (k + \frac{1}{2})h_S(x) + \frac{1}{2}h_T(x)$$

is an unit interval representation of S using integer endpoints.

Theorem 5. *The value d is the minimal possible length for intervals in an unit interval representation with integer endpoints.*

As an example we show in Figure 4 the minimal unit interval representation for the semiorder of Figure 3. We obtain in this example, $d = 1/2(h_T(z_1) - h_T(a_1)) = 4$ and therefore a representation by $l(x) = 5h_S(x) + 1/2(h_T(x) - h_S(x))$. Note that, because of Proposition 3, we have

$$d = 1/2 \max\{h_T(x) - h_T(y) : x, y \in X \text{ and } h_S(x) = h_S(y)\} .$$

The proof of Theorem 4 will be made in several steps. First, we have to show that the mapping is indeed an interval representation for every $k \geq d$, i.e.,

$$x <_S y \iff l(x) + k < l(y) .$$

This will be done in the Lemmas 6, 7 and 8. Subsequently, we will prove Theorem 5.

Lemma 6. 1. $(a, b) \in C \Longrightarrow l(b) > l(a) + k$.
2. $(b, a) \in I \Longrightarrow l(a) \geq l(b) - k$.

Proof. 1. If $(a, b) \in C$ then $h_S(b) - h_S(a) = 1$ by Lemma 1, and $h_T(b) - h_T(a) \geq 1$ by the definition of T. Therefore $l(b) - l(a) = (k + 1/2)(h_S(b) - h_S(a)) + 1/2(h_T(b) - h_T(a)) \geq k + 1/2 + 1/2 = k + 1$, as desired.

2. $(b, a) \in I^*$ implies $h_S(b) - h_S(a) = 1$ and $h_T(b) - h_T(a) \leq -1$. Therefore $l(b) - l(a) \leq k + 1/2 - 1/2 = k$. If $(b, a) \in I \setminus I^*$ then $h_S(b) = h_S(a)$ and $h_T(b) - h_T(a) \leq 2d \leq 2k$. We obtain likewise $l(b) - l(a) \leq k$. □

Lemma 7. $x \leq_L y \Longrightarrow l(x) \leq l(y)$.

Proof. For $x <_L y$ with $h_S(x) = h_S(y)$ the claim follows easily with Proposition 3. It remains to show $l(z_i) < l(a_{i+1})$ for $i \in \{0, \ldots, h(S) - 1\}$. Let $u := \max_L Pred(a_{i+1})$. Then $(u, a_{i+1}) \in C$. It follows $h_S(u) = i$ and therefore $l(z_i) \leq l(u) + d \leq l(u) + k < l(a_{i+1})$. □

Lemma 8. $x <_S y \iff l(x) + k < l(y)$.

Proof. Let $x <_S y$. Then define $a := \max_L Pred(y)$ and $b := \min_L Suc(a)$. It follows $x \leq_L a$, $b \leq_L y$ and aCb. With Lemma 6 and 7 we obtain $l(x) + d \leq l(a) + k < l(b) \leq l(y)$, as desired.

Now, let $x \not<_S y$. Then let further $b := \max_L X \setminus Suc(x)$ and $a := \min_L X \setminus Pred(b)$. As consequence we get $y \leq_L b$, bIa and $a \leq_L x$. It follows $l(y) \leq l(b) \leq l(a) + k \leq l(x) + k$, and the claim is proven. □

We have shown that the mapping $x \longmapsto [l(x), l(x)+k]$, for every $k \geq d$, is an unit interval representation of S. Note that we use certainly integer endpoints in this representation because of the structure of T: Each path in T between two elements of the same level has an even length and each path between two consecutive levels has an odd length. Since moreover T has only one minimal element, the height of an element is even in T if and only if it is even in S. Hence d is an integer and, as desired, $l(x)$ is an integer for every $x \in X$.

To prove the minimality of d we will use an idea which is based on a correspondence between the representation of semiorders and valued graphs described by Pirlot [10]. He considers the graph on X with the arc set $C \cup I$ where he assigns to the arcs of C the value $k+1$ and to the arcs of I the value $-k$, for some $k \in \mathbb{N}$. The reason is that for any unit interval representation $x \longmapsto [f(x), f(x)+k]$ of S we have

$$(x, y) \in C \Longrightarrow x <_S y \Longrightarrow f(y) \geq f(x) + k + 1 \quad (1)$$

$$(y, z) \in I \Longrightarrow z \not<_S y \Longrightarrow f(z) \geq f(y) - k \quad (2)$$

Obviously, a function f satisfying (1) and (2) can be obtained only if there is no cycle in $C \cup I$ of strictly positive value.

Proof of Theorem 5. Let $x \longmapsto [f(x), f(x)+k]$ be an arbitrary interval representation of S with integer endpoints. Further let $i \in \{0, \ldots, h(S)\}$ such that $h_T(z_i) - h_T(a_i) = 2d$. Then there is a path of length $2d$ in T leading from a_i to z_i. This path contains d arcs of C and d arcs of I^*. Furthermore, we have $z_i \not<_S a_i$ and even $(z_i, a_i) \in I$ since the path in T between a_i and z_i is of maximal length for elements of the same height. Therefore there is a cycle in $C \cup I$ with d arcs of value $k+1$ and $d+1$ arcs of value $-k$. The value of the cycle is therefore $d(k+1) - (d+1)k = d - k$. The value may not be positive. Hence $k \geq d$, i.e., every unit interval representation of S with integer endpoints has interval length greater or equal than d. □

5 A Linear Algorithm

In the following we will describe how to obtain an algorithm with time and space complexity $O(|V|+|E|)$ which, for a given directed graph $G = (V, E)$, determines first if its transitive closure G^{tc} is a semiorder, and secondly computes, in the affirmative case, the minimal unit interval representation (as given in Theorem 4 and 5) of this semiorder.

The algorithm consists of the following steps:

1. Recognition if G^{tc} is an interval order and calculation of the canonical interval representation (Note: This is in general not a *unit* interval representation).

2. Checking if G^{tc} is a semiorder and computation of a canonical linear extension L of $S := G^{tc}$.
3. Calculation of the height $h_S(x)$ of each element in S and of the minimal and maximal elements of each level of S with respect to L, denoted by a_i and z_i.
4. Calculation of the arcs sets C and I^*.
5. Calculation of the height $h_T(x)$ of the elements in T and of the minimal interval length d.
6. Calculation of the intervals.

We will describe the different steps more precisely in the proof of the following theorem.

Theorem 9. *The previous algorithm recognizes for a given directed graph $G = (V, E)$, if G^{tc} is a semiorder, and in this case, computes a minimal unit interval representation. The time and space complexity of the algorithm is $O(|V| + |E|)$.*

Proof. 1. The recognition of interval orders can be done with an algorithm by Baldy and Morvan [2] which has $O(|V| + |E|)$ time and space complexity. The algorithm produces also the canonical interval representation as defined in Section 1, i.e., an element $x \in V$ gets the interval $[l(x), r(x)]$ where $l(x) = |\{Pred(y) : Pred(y) \subset Pred(x)\}|$ and $r(x) = |\{Suc(y) : Suc(y) \supset Suc(x)\}|$.

2. Calculate $g(x) := l(x) + r(x)$ for each $x \in V$, and $m := \max\{l(x)|x \in V\}$ $(= \max\{r(x)|x \in V\})$. Note that $m \leq |V|$. Sort V according to the mapping g with a bucket sort using $2(m + 1)$ cells. The verification that G^{tc} is a semiorder and the computation of the canonical linear extension L for the (reduced) semiorder can now be carried out as follows: For each nonempty cell compare the intervals of its elements. There must be equal if G^{tc} is a semiorder. Choose a representative of each cell. The canonical linear extension L will be obtained by taking the representatives of the nonempty cells in increasing order. For two consecutive elements x_i, x_{i+1} in L verify if $l(x_i) \leq l(x_{i+1})$; otherwise G^{tc} is not a semiorder.

3. Let x_1 be the minimal element of L. Then $h_S(x_1) = 0$ and $a_0 := x_1$. Set $h := r(a_0) + 1$. The following elements in L have height 0 until we reach the first element, denoted by x_j, with $l(x_j) = h$. Then $h_S(x_j) = 1$, $z_0 := x_{j-1}$, and $a_1 := x_j$. Set now $h = r(a_1) + 1$. Continue this procedure until the last element of L, x_n, is reached.

4. The arcs of C will be stored in 2 one-dimensional arrays C^+ and C^- such that the source of the i-th arc of C is stored in $C^+(i)$ and the sink is stored in $C^-(i)$. Analogously two arrays I^+ and I^- will be generated. Note that

$$C = \{(\max_L\{x : r(x) = i\}, \min_L\{x : l(x) = i + 1\}) : 0 \leq i \leq m - 1\} \ ,$$
$$I = \{(\max_L\{x : l(x) = i\}, \min_L\{x : r(x) = i\}) : 0 \leq i \leq m\} \ .$$

Hence $|C| = m$ and $|I| = m + 1$. The computation of the arrays can be done as follows:

$I^-(0) := x_1;\;\; I^+(m) := x_n$
For $i = 1$ **to** $n-1$ **do**
If $l(x_i) < l(x_{i+1})$ **then** $I^+(l(x_i)) := x_i$; $C^-(l(x_{i+1})-1) := x_{i+1}$
If $r(x_i) < r(x_{i+1})$ **then** $C^+(r(x_i)) := x_i$; $I^-(r(x_{i+1})) := x_{i+1}$

Now remove the elements not belonging to I^* by setting $I^+(i) = I^-(i) = -1$ if $h_S(I^+(i)) = h_S(I^-(i))$.

5. The complexity of a standard algorithm to compute the height of the elements in a directed graph is linear in the number of elements and arcs. To compute the height in T it suffices to consider the arcs of the transitive reduction of T, i.e., the arcs of $C \cup I^*$. Since $|C \cup I^*| \leq 2m+1$ and $m \leq |V|$, the calculation can be done in $O(|V|)$. Now, the calculation of $d = 1/2 \max\{h_T(z_i) - h_T(a_i) : 0 \leq i \leq h(S)\}$ is easy.

6. Likewise, the computation of the intervals is immediate.

It is clear that the steps 1 to 6 require linear time and space complexity. □

Certainly, in the above algorithm, some of the steps could be executed at the same time. But for a matter of clearness, we preferred to describe the algorithm in several separate steps.

6 Forbidden Suborders

Let $\mathcal{S}_k$ be the subclass of semiorders that can be represented by an unit interval representation with integers and intervals of length k. Note that from Theorem 4 it follows

$$\mathcal{S}_0 \subseteq \mathcal{S}_1 \subseteq \mathcal{S}_2 \subseteq \ldots$$

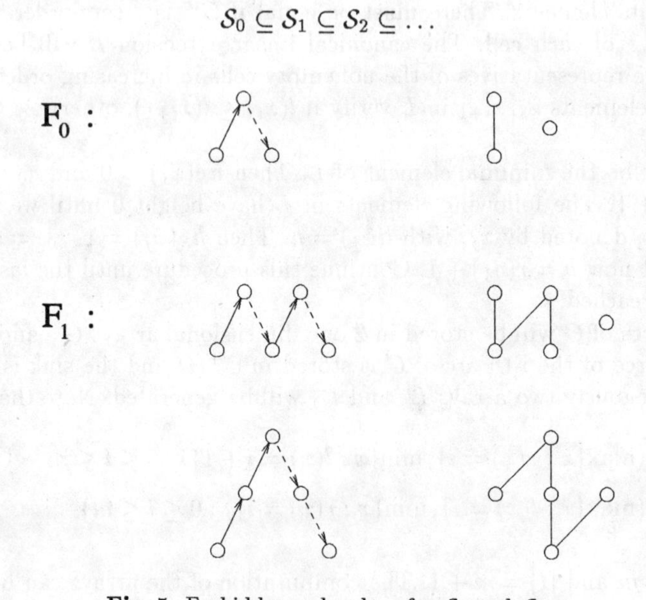

Fig. 5. Forbidden suborders for $\mathcal{S}_0$ and $\mathcal{S}_1$.

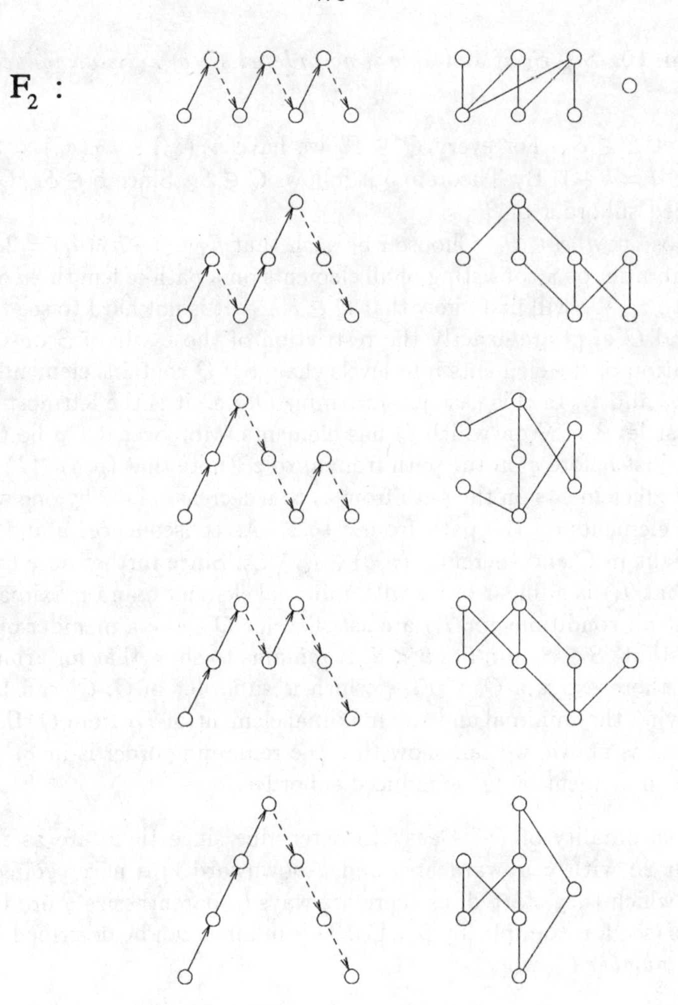

Fig. 6. Forbidden suborders for $\mathcal{S}_2$.

We will characterize $\mathcal{S}_k$ by a set F_k of minimal forbidden suborders (cf. also the paper by Bogart and Stellpflug [4]). The elements of F_k can be totally determined by means of the structure of their associated order T as defined in Section 3. Any element of F_k is a semiorder Q with $2k+3$ elements where the associated order T_Q is a linear order and the maximum of T_Q has height 0 in Q, i.e., it lies on the lowest level of Q. For $k \geq 0$, $k \in \mathbb{N}$, let us formally define

$$F_k := \{Q = (X, <_Q) \text{ semiorder} : |X| = 2k+3, T_Q \text{ linear and } h_Q(\max T_Q) = 0\} \ .$$

Thus for every $Q \in F_k$ the order T_Q contains $k+1$ arcs of C and $k+1$ arcs of I^*. In the Figures 5 and 6 all orders of F_0, F_1 and F_2 are depicted, first showing the order T_Q and then the Hasse diagram of Q.

Theorem 10. *$S \in \mathcal{S}_k$ if and only if no ordered set of F_k is an induced suborder of S.*

Proof. Let $S \in \mathcal{S}_k$. For every $Q \in F_k$ we have $h_T(z_0) - h_T(a_0) = 2k + 2$ and therefore $d = k + 1$. By Theorem 5 it follows $Q \notin \mathcal{S}_k$. Since $S \in \mathcal{S}_k$, Q cannot be an induced suborder of S.

Suppose now, $S \notin \mathcal{S}_k$. Choose i be such that $h_T(z_i) - h_T(a_i) = 2d$ and let Q be the suborder of S consisting of all elements on a path of length $2d$ of T leading from a_i to z_i. We will first show that $Q \in F_{d-1}$. It is not hard to see that the arc sets I and C of Q are exactly the restriction of those sets of S on Q. However the partition of the elements into levels change if Q contains elements below the level of a_i and z_i. In this case let $q := \min_L Q$, i.e., it is the leftmost element on the lowest level of S on which Q has elements. Moreover, let p be the element occuring just before q on the path from a_i to z_i (note that $(p, q) \in I_S^*$). Then the height of all elements on the path from a_i to p decreases in Q by one with respect to those elements on the path from q to z_i. As consequence, p and q have the same height in Q and therefore $(p, q) \in I_Q \setminus I_Q^*$. Since furthermore $(z_i, a_i) \in I_Q^*$, we get that T_Q is a linear order with minimal element q and maximal element p such that all conditions for T_Q are satisfied for Q to be a member of F_{d-1}.

Note that, $S \notin \mathcal{S}_k$ implies $k < d$. It remains to show that for arbitrary j and $Q \in F_j$, there exists a $Q' \in F_{j-1}$ which is suborder of Q. Q' can be obtained by removing the minimal and the maximal element of T_Q from Q. Using similar arguments as above, we can show that the remaining order is in F_{j-1}. Hence S contains an element of F_k as induced suborder. □

The cardinality of F_k is easy to determine since there are as many paths of length $2k$ with k upward arcs and k downward arcs never going below the level on which they started, as there are ways to parenthesize a product of $k + 1$ variables (see for example [8], p. 345). This number can be described by the k–th *Catalan number* C_k where

$$C_k := \frac{1}{k+1}\binom{2k}{k} .$$

Therefore we have

Corollary 11. $|F_k| = C_{k+1}$.

Recall that there is another correspondence between semiorders and Catalan numbers:

Theorem 12 Wine and Freund [14]. *The number of unlabeled semiorders on k elements is C_k.*

References

1. A. von Arnim and C. de la Higuera. Computing the jump number on semi-orders is polynomial. *Discrete Applied Mathematics*, 1993. To appear.

2. P. Baldy and M. Morvan. A linear time and space algorithm to recognize interval orders. *Discrete Applied Mathematics*, 46:173–178, 1993.
3. K. P. Bogart. A discrete proof of the Scott–Suppes representation theorem of semiorders. Technical Report PMA–TR91–173, Dartmouth College, Hanover, New Hampshire 03755, 1991. Accepted for publication in *Discrete Applied Mathematics*.
4. K. P. Bogart and K. Stellpflug. Discrete representation theory of semiorders. Accepted for publication in *Discrete Applied Mathematics*.
5. G. Brightwell. Semiorders and the 1/3–2/3 conjecture. *Order*, 5:369–380, 1989.
6. G. Brightwell and P. Winkler. Counting linear extensions is #P–complete. In *Proc. 23rd ACM Symposium on the Theory of Computing*, pages 175–181, 1991.
7. P.C. Fishburn. *Interval orders and interval graphs.* Wiley, New York, 1985.
8. R. L. Graham, D. E. Knuth, and O. Patashnik. *Concrete Mathematics: a foundation of computer science.* Addison–Wesley Publishing Company, Reading, Massachusetts, 1989.
9. J. Mitas. Tackling the jump number of interval orders. *Order*, 8:115–132, 1991.
10. M. Pirlot. Minimal representation of a semiorder. *Theory and Decision*, 28:109–141, 1990.
11. I. Rabinovitch. The dimension of semiorders. *J. of Combinatorial Theory A*, 25:50–61, 1978.
12. I. Rabinovitch. An upper bound on the dimension of interval orders. *J. of Combinatorial Theory A*, 25:68–71, 1978.
13. D. Scott and P. Suppes. Foundational aspects of theories of measurement. *Journal of Symbolic Logic*, 23:113–128, 1958.
14. R. L. Wine and J. E. Freund. On the enumeration of decision patterns involving n means. *Annals of Mathematical Statistics*, 28:256–259, 1957.

The Computation of the Jump Number of Convex Graphs

Elias Dahlhaus *

Basser Dept. of Computer Science, University of Sydney, NSW 2006, Australia

Abstract: A first polynomial time algorithm for the computation of the jump number of a convex bipartite graph is presented. The algorithm uses dynamic programming methods.

Introduction

A bipartite graph $(V_1 \cup V_2, E)$ is called *convex* iff the vertices of V_1 can be enumerated in a way, say $V_1 = \{v_1, \ldots, v_t\}$, such that for each $w \in V_2$, the set $N(w) = \{v \in V_1 : wv \in E\}$ forms an interval $\{v_j, v_{j+1}, \ldots, v_k\}$. It is well known that the jump number problem for bipartite graphs and therefore for convex bipartite graphs is equivalent to the problem of finding a maximum matching, having no alternating cycles [3].

The jump number problem is NP-complete for chordal bipartite graphs [4] and solvable in polynomial time for biconvex graphs [1]. The jump number problem for convex bipartite graphs was open for a long time.

Here we give a polynomial time algorithm based on dynamic programming, which computes an alternating cycle free matching of maximal cardinality. The exponent of the polynom is quiet high. It might still be interesting to get a better polynomial time algorithm.

In section 1 we introduce the necessary notation. In section 2, we develop the algorithm. In section 3 we present a small improvement of the algorithm. Section 4 gives a short overview on possible future research.

1 Basic Notions

A graph consists here of a set of *vertices* and *edges*. Multiple edges and loops are not allowed. A bipartite graph is a graph consisting of a vertex set $V_1 \cup V_2$ of disjoint sets V_1 and V_2, such that for all edges xy, we have $x \in V_1$ and $y \in V_2$ or vice versa.

A bipartite graph $(V_1 \cup V_2, E)$ is called convex iff

1. V_1 is a set of integers,
2. V_2 is a set of intervals, and
3. for $x \in V_1$ and $I \in V_2$, $xI \in E$ iff $x \in I$.

One ever can interpret a bipartite graph $(V_1 \cup V_2, E)$ as a partial ordering $<$ in the canonical way:

$$x < y \Leftrightarrow x \in V_1,\, y \in V_2,\, xy \in E.$$

---

* Part of the work has been done, while the author visited the Department of Mathematics of the Friedrich-Schiller-University of Jena in March 1990. Most of the work has been done while the author was with the Computer Science Department of the University of Bonn

For any extension $<'$ of $<$ to a total ordering and its associated enumeration $\{x_i\}_{i=1}^n$, let (x_i, x_{i+1}) be a *jump* iff $x_i \not< x_{i+1}$.

The *jump number problem* is to find an extension of the partial order $<$ to a total ordering with a minimum number of jumps.

To find an equivalent formulation, we introduce the notion of a matching. A *matching* in $G = (V, E)$ is a subset M of E, such that no two edges of M have a common vertex. An alternating cycle of M consists of a cycle $(e_1, e_2, e_3, \ldots, e_{2n}, e_1)$ such that $e_{2j} \in M$ and $e_{2j-1} \notin M$ for $j = 1, \ldots, n$.

Proposition: [3] *For bipartite graphs, the jump number problem is polynomially equivalent to the problem of finding a maximum alternating cycle free matching.*

Proposition: [4]*For chordal bipartite graphs and therefore for convex bipartite graphs, any matching, which has no alternating cycle of length four, has no alternating cycle.*

Therefore we have to find a matching, which has no alternating cycle of length four and is of maximal cardinality.

The jump number problem for convex bipartite graphs can also be stated as follows:

Given a set $\mathcal{I}$ of intervals and a set N of numbers, find a maximum set M of pairs (x, I), such that

1. $x \in N$,
2. $x \in I$,
3. for no $(x_1, I_1), (x_2, I_2) \in M$, we have $x_1 \in I_2$ and $x_2 \in I_1$.

The last condition states that there is no alternating cycle of length four.

2 The Development of the Algorithm

The following situation is an indicator that simple left-right scanning algorithms cannot solve the problem to find a maximum alternating cycle free matching in a convex bipartite graph.

$I_1 \subset I_2$, $x \in I_2 \setminus I_1$, and $y \in I_1$.

Then $\{(x, I_2), (y, I_1)\}$ still forms an alternating cycle free matching.

If in an alternating cycle free matching such a situation appears, such that $x < y$, then it is not possible, to determine such a matching by a simple left to right scanning procedure.

To get such a situation in control, we introduce the notion of a *matching tree*:

Given an alternating cycle free matching $M = \{(x_1, I_1), \ldots, (x_k, I_k)\}$, such that $x_i < x_j$, for $i < j$. Then we can define the following tree T by the parent function P:

For each (x_i, I_i), let $P(x_i, I_i)$ be the (x_j, I_j), such that

1. $I_i \subset I_j$,
2. $x_j < x_i$, and
3. j is maximal under these conditions.

If no such (x_j, I_j) exists, then $P(x_i, I_i) = U$, where U is the root of the matching tree and stands for "undefined".

Since $I_i \subset I_j$, P defines a parent function of a tree.

Moreover $x_i \notin I_j$. Otherwise an alternating cycle of length four would exist.

The following observation is an immediate consequence of the definition of P.

Lemma 1: *Suppose (x_i, I_i) and (x_j, I_j) are the children of the same parent (x_k, I_k), i.e. $P(x_i, I_i) = P(x_j, I_j) = (x_k, I_k)$.*
Then the following situation cannot happen:

$$x_i < x_j \text{ and } I_j \subset I_i.$$

Next we prove, that the set of descendents of (x_i, I_i) is the set of (x_j, I_j), such that $x_i < x_j \leq x_p$, for some p. This follows inductively from the following result:

Lemma 2: *Suppose $(x_i, I_i) = P(x_j, I_j)$ and $x_i < x_k < x_j$.*
Then $P(x_k, I_k)$ is a (x_p, I_p), such that
$x_i \leq x_p < x_k$ and $I_p \subseteq I_i$.

Proof: Since $(x_i, I_i) = P(x_j, I_j)$, $x_j \in I_i$ and therefore $x_k \in I_i$. Since $x_k \in I_i$, $x_i \notin I_k$ otherwise an alternating cycle of length four would be created). Therefore the left border of I_k is greater than x_i.

Because of the maximality of x_i such that I_i contains I_j, I_k cannot contain I_j. Note that $x_j \notin I_k$ or $x_k \notin I_j$. Therefore the right border of I_k is smaller than the right border of I_j and therefore smaller than the right border of I_i. hereby $I_k \subset I_i$.

Therefore the parent of (x_k, I_k) must be some (x_p, I_p), such that $x_i \leq x_p$.

If $x_i = x_p$, then $P(x_k, I_k) = (x_i, I_i)$ and we are done.

If $x_i \neq x_p$, then $x_i < x_p$. Then, by the same arguments, as for x_k and I_k, the right border of I_p is smaller than the right border of I_j and the left border of I_p is greater than x_i. Therefore $I_p \subset I_i$.

□

This lemma implies the following result:

Proposition 3: *Suppose (x_j, I_j) is a descendent of (x_i, I_i), that means (x_i, I_i) arises from (x_j, I_j) by iterated application of P. Suppose $x_i < x_k < x_j$.*
Then (x_k, I_k) is a descendent of (x_i, I_i).

Proof: Suppose $P^l(x_j, I_j) = (x_i, I_i)$. That means the l-times application of P to (x_j, I_j) is (x_i, I_i). We prove this proposition by induction on l.

Let $l = 1$: Then by lemma 2 $P(x_k, I_k)$ is some (x_p, I_p), such that $x_i \leq x_p < x_k$ and $I_p \subseteq I_i$. This x_p is again an x_k satisfying the assumptions of lemma 2. Since we have only a finite number of pairs (x_m, I_m), by iterated application of P we reach (x_i, I_i).

The inductive step from l to $l + 1$ can be done as follows:

Suppose $P(x_j, I_j) = (x_p, I_p)$. Moreover we assume $P^l(x_p, I_p) = (x_i, I_i)$. Then we have to consider the following cases:

1. $x_k < x_p$. Then by the assumption of the induction, (x_k, I_k) is a descendent of (x_i, I_i).
2. $x_k > x_p$. Then we only have to consider the case that $l = 1$ and we can conclude that (x_k, I_k) is a descendent of (x_p, I_p). Therefore (x_k, I_k) is a descendent of (x_i, I_i).

□

We consider conditions, when two alternating cycle free matchings

$$\{(x_1, I_1), \ldots, (x_k, I_k)\}$$

and

$$\{(x_{k+1}, I_{k+1}), \ldots, (x_l, I_l)\}$$

can be united to one alternating cycle free matching, provided that $x_1 < x_2 < \ldots < x_k < x_{k+1} < \ldots < x_l$.

Suppose both matchings are unions of descendent sets of some (x_i, I_i), having the same parent. That means (x_1, I_1) and (x_{k+1}, I_{k+1}) have the same parent and no descendent of (x_{k+1}, I_{k+1}) belongs to the first matching. Then for each $j = 1, \ldots, k$,

$$I_{k+1} \not\subset I_j.$$

That is the reason why it is appropriate to deal with the question when two alternating cycle free matchings $\{(x_1, I_1), \ldots, (x_k, I_k)\}$ and $\{(x_{k+1}, I_{k+1}), \ldots, (x_l, I_l)\}$ can be concatenated, provided $I_{k+1} \not\subset I_j$, for each $j = 1, \ldots, k$. We can prove the following result:

Proposition 4: *Suppose $\{(x_1, I_1), \ldots, (x_k, I_k)\}$ and $\{(x_{k+1}, I_{k+1}), \ldots, (x_l, I_l)\}$ are alternating cycle free matchings, $x_1 < \ldots < x_k < x_{k+1} < \ldots x_l$, and I_{k+1} is not a subset of any I_j with $j \leq k$.*

Then $\{(x_1, I_1), \ldots, (x_l, I_l)\}$ is an alternating cycle free matching iff $\{(x_1, I_1), \ldots, (x_{k+1}, I_{k+1})\}$ is an alternating cycle free matching.

Proof: $\Rightarrow$: trivial

$\Leftarrow$: We only have to check, that no alternating cycle of length four exists. That means, for no pair $(x_i, I_i), (x_j, I_j)$ with $x_i < x_j$, we have $x_i \in I_j$ and $x_j \in I_i$. We only have to check it for $i \leq k$ and $j > k + 1$.

We consider two cases:

1. *Suppose* $x_j \notin I_{k+1}$: Then the right border of I_{k+1} is smaller than x_j. Since $I_{k+1} \not\subset I_i$, for any $i \leq k$, the right border of I_{k+1} is larger than the right border of any I_i with $i \leq k$. Therefore $x_j \notin I_i$, for any $i \leq k$. Therefore x_j cannot touch any alternating cycle of length four.
2. *Suppose* $x_j \in I_{k+1}$. Then x_{k+1} does not contain I_j. Otherwise an alternating cycle in $\{(x_{k+1}, I_{k+1}), \ldots x_l\}$ would exist. But then I_j does not contain any x_i with $i \leq k$. Therefore $(x_j, I_j$ does not appear in any alternating cycle of length four.

□

Now we consider necessary and sufficient conditions, when $\{(x_1, I_1), \ldots, (x_k, I_k), (x_{k+1}, I_{k+1})\}$ is an alternating cycle free matching.

We assume that children (x, I) of the same parent are ordered with respect to x from left to right. For this purpose, we consider the *right most sequence* $((x_{i_1}, I_{i_1}), \ldots, (x_{i_p}, I_{i_p}))$ of $M = \{(x_1, I_1), \ldots, (x_k, I_k)\}$. It consists of the sequence of the right most path of the matching tree of $\{(x_1, I_1), \ldots, (x_k, I_k)\}$.

Lemma 5: *For each $i \leq k$, there is an (x_j, I_j) of the right most sequence of M, such that $x_i \leq x_j$ and the right border of I_i is at most as large as the right border of I_j.*

Proof: We may assume that (x_i, I_i) does not appear in the right most sequence of M. If (x_i, I_i) is not a descendent of the first element (x_{i_1}, I_{i_1}) of the right most sequence of M then by proposition 3, $x_i < x_{i_1}$. Since the parent of (x_{i_1}, I_{i_1}) is U, I_{i_1} is not a subset of I_i. Note that, since $x_i < x_{i_1}$ and M has no alternating cycle of length four, i.e I_i does not contain x_{i_1} of I_{i_1} does not contain x_i, the left border of I_i is less than the left border of I_{i_1}. Therefore the right border of I_i is less than the right border of I_{i_1}.

Now we assume that (x_i, I_i) is a descendent of (x_{i_j}, I_{i_j}) but not a descendent of $(x_{i_{j+1}}, I_{i_{j+1}})$. By proposition 3, $x_i < x_{i_{j+1}}$. Moreover, (x_i, I_i) is a descendent of some $(x_{i'}, I_{i'})$ with parent (x_{i_j}, I_{i_j}) ($(x_{i_{j+1}}, I_{i_{j+1}})$ has the same parent). By lemma 1, $I_{i_{j+1}}$ is not a subset of $I_{i'}$ and therefore also not a subset of I_i (which is a subset of $I_{i'}$). By the same argument as in the case that (x_i, I_i) is not a descendent of (x_{i_1}, I_{i_1}), the left border of I_i is less than the left border of $I_{i_{j+1}}$ and the right border of I_i is less than the right border of $I_{i_{j+1}}$.

□

Now we have a criterion when (x_{k+1}, I_{k+1}) can be added.

Proposition 6: *Suppose* $M = \{(x_1, I_1), \ldots (x_k, I_k)\}$ *is an alternating cycle free matching,* $x_1 < \ldots < x_k < x_{k+1}$*, and* $I_{k+1} \not\subset I_j$*, for* $j \leq k$*. Denote the right most sequence of M by* $((x_{i_1}, I_{i_1}) \ldots, (x_{i_q}, I_{i_q}))$*. Then* $\{(x_1, I_1), \ldots, (x_k, I_k), (x_{k+1}, I_{k+1})\}$ *is an alternating cycle free matching iff* $x_{k+1} \notin I_{i_1}$ *or* $x_{i_q} \notin I_{k+1}$ *or there is a right most sequence element* (x_{i_j}, I_{i_j})*, such that*

$$x_{i_j} \notin I_{k+1}$$

and

$$x_{k+1} \notin I_{i_{j+1}}.$$

Proof: ⇒: Suppose $\{(x_1, I_1), \ldots, (x_{k+1}, I_{k+1})\}$ is alternating cycle free. We have to consider three cases:

1. *Suppose* I_{k+1} *contains all* x_{i_j}. Then x_{k+1} is in no I_{i_j} and therefore not in I_{i_1}. Otherwise an alternating cycle of length four would exist.
2. *Suppose* $x_{k+1} \in I_{i_j}$*, for all* I_{i_j}. Then I_{k+1} cannot contain any x_{i_j} and therefore $x_{i_q} \notin I_{k+1}$.
3. *It remains to consider the following situation*:

$$x_{i_j} \notin I_{k+1} \text{ and } x_{i_{j+1}} \in I_{k+1}.$$

Then $x_{k+1} \notin I_{i_{j+1}}$. Otherwise an alternating cycle of length four would exist.

⇐: First we show that (x_{k+1}, I_{k+1}) does not create an alternating cycle with any (x_{i_j}, I_{i_j}).

Suppose $x_{i_j} \notin I_{k+1}$. Then also $x_{i_{j'}} \notin I_{k+1}$, for any $j' < j$, because in that case $x_{j'} < x_j$.

We consider three possible cases:

1. Suppose $x_{k+1} \notin I_{i_1}$. Then $x_{k+1} \notin I_{i_j}$, for any j, since $I_{i_j} \subset I_{i_1}$. Therefore an alternating cycle with any right most sequence element cannot be created.

2. Suppose $x_{i_q} \notin I_{k+1}$. Then for any j, $x_{i_j} \notin I_{k+1}$. Therefore an alternating cycle of length four with any right most sequence element cannot be created.
3. Suppose $x_{i_j} \notin I_{k+1}$ and $x_{k+1} \notin I_{i_{j+1}}$. Then for any $j' \leq j$ $x_{i_{j'}} \notin I_{k+1}$ and for any $j' > j$, $x_{k+1} \notin I_{i_{j'}}$. Therefore (x_{k+1}, I_{k+1}) cannot form an alternating cycle of length four with any $(x_{i_{j'}}, I_{i_{j'}})$.

Consider now any $i \leq k$. Then we find some $x_{i_j} \geq x_i$ with the property that the right border of I_{i_j} is at least the right border of I_i.

If (x_i, I_i) forms an alternating cycle with (x_{k+1}, I_{k+1}) then $x_{k+1} \in I_i$. In that case also $x_{k+1} \in I_{i_j}$. But then $x_{i_j} \notin I_{k+1}$, because (x_{k+1}, I_{k+1}) does not form an alternating cycle with any right most sequence element. But then also $x_i \notin I_{k+1}$. Therefore (x_i, I_i) cannot form an alternating cycle with (x_{k+1}, I_{k+1}).

□

Next we observe the following:

Lemma 7: *Suppose $\{(x_1, I_1), \ldots, (x_l, I_l)\}$ is an alternating cycle free matching and $I_{k+1} \not\subset I_i$, for any $i \leq k$.*

Then the right most sequence of $\{(x_1, I_1), \ldots, (x_l, I_l)\}$ coincides with the right most sequence of $\{(x_{k+1}, I_{k+1}), \ldots, (x_l, I_l)\}$.

Proof: Since $I_{k+1} \not\subset I_i$, for any $i \leq k$, the right border of I_{k+1} is at least the right border of any I_i with $i \leq k$. Therefore the I_j, such that the right border is maximal, has the property that $j \geq k+1$. Therefore the right most sequence of $\{(x_1, I_1), \ldots, (x_l, I_l)\}$, denoted by $((x_{i_1}, I_{i_1}), \ldots, (x_{i_q}, I_{i_q}))$ has the property that $x_{k+1} \leq x_{i_1} < \ldots < x_{i_q}$. Therefore the right most sequence of $\{(x_1, I_1), \ldots, (x_l, I_l)\}$ is fully contained in $\{(x_{k+1}, I_{k+1}), \ldots, (x_l, I_l)\}$. Therefore the right most sequences of these both matchings coincide.

□

For each alternating cycle free matching $M = \{(x_1, I_1), \ldots, (x_k, I_k)\}$, we consider the following data:

$x_1^M = x_1$, $x_2^M = x_k$, $[u_M, v_M] = I_1$, the left most left border v_{min}^M of some I_j, the right most right border v_{max}^M of some I_j, and for the right most sequence $((x_{i_1}, I_{i_1}), \ldots, (x_{i_q}, I_{i_q}))$, one pair $r_j^M = x_{i_j}$, $s_j^M =$ right border of $I_{i_{j+1}}$, if $j < q$ or $s_j^M = -\infty$ if $j = q$.

We call $(x_1^M, x_2^M, u_M, v_M, v_{min}^M, v_{max}^M, r_j^M, s_j^M, \sharp M)$ an *admissible tuple* for M. Such a tuple is called admissible if there is an alternating cycle free matching M for which it is admissible.

The number of admissible tuples is polynomially bounded by the number of intervals and points. Our aim is now to develop an algorithm computing all admissible tuples. From this we get an alternating cycle free matching of maximum cardinality.

We generate the set of admissible tuples as follows:

Initialization If $I = [u, v] \in \mathcal{I}$ and $x \in I$, then

$$(x, x, u, v, u, v, x, -\infty, 1)$$

is an admissible tuple of the one element matching (x, I) and the only one.

Rule 1 If $(x_1, x_2, u, v, v_{min}, v_{max}, r, s, m)$ and $(\bar{x}_1, \bar{x}_2, \bar{u}, \bar{v}, \bar{v}_{min}, \bar{v}_{max}, \bar{r}, \bar{s}, \bar{m})$ are admissible and the following statements are valid:

$$x_2 < \bar{x}_1, v_{max} < \bar{v}, \text{ and } v_{max} < x_1 \text{ or } r < \bar{u}\, and\, s < \bar{x}_1,$$

then

$$(x_1, \bar{x}_2, u, v, \min(v_{min}, \bar{v}_{min}), \bar{v}_{max}, \bar{r}, \bar{s}, m + \bar{m})$$

is admissible.

This follows from proposition 4 and proposition 6.

Rule 2 Suppose that $(x_1, x_2, u, v, v_{min}, v_{max}, r, s, m)$ is admissible, $x \in I = [\tilde{u}, \tilde{v}] \in \mathcal{I}$, and that the following statements are true:

$$x < x_1, x < v_{min}, v_{max} < \tilde{v}.$$

Then

$$(x, x_2, \tilde{u}, \tilde{v}, \tilde{u}, \tilde{v}, r, s, m + 1)$$

and

$$(x, x_2, \tilde{u}, \tilde{v}, \tilde{u}, \tilde{v}, x, v_{max}, m + 1)$$

are admissible.

The correctness of rule 2 can be seen as follows:

Consider any alternating cycle free matching $M = \{(y_1, I_1), \ldots (y_k, I_m)\}$, for which $(x_1, x_2, u, v, v_{min}, v_{max}, r, s, m)$ is admissible.

Then $x_1 = y_1$, $x_2 = y_m$, $[u, v] = I_1$, v_{max} is the right most right border, and v_{min} is the left most left border of some I_j.
Let $((y_{i_j}, I_{i_j}))_{j=1}^q$ be the right most sequence of M. Then r is some y_{i_j} and s is the right border of $I_{i_{j+1}}$, if $j \neq q$, otherwise $s = -\infty$. Moreover v_{max} is the right border of I_{i_1}.

The following statements can immediately be derived:

- (x, I) can be added not destroying the alternating cycle freeness, because $x \notin I_i$, for any i.
- The right most sequence of

$$\{(x, I), (y_1, I_1), \ldots, (y_m, I_m)\}$$

is

$$((x, I), (y_{i_1}, I_{i_1}), \ldots (y_{i_q}, I_{i_q})).$$

From these two facts it follows, that the two tuples generated by rule 2 are admissible.

Therefore:

Proposition 8: *All tuples derivable from the initialization by rules 1 and 2 are admissible.*

□

Vice versa we also have to prove:

Proposition 9: *All admissible tuples can be derived from the initialization by successive application of rule 1 and rule 2.*

Proof: We consider any alternating cycle free matching $M = \{(x_1, I_1), \ldots (x_k, I_k)\}$ and its matching tree T with its parent function P. We shall prove by induction on the size of M, that each admissible tuple for M is derivable by the above rules. We consider three cases:

1. If the size of M is one, then each admissible tuple for M is derivable by the initialization rule.
2. Suppose I_1 covers all other I_i. Then x_1 is not in any I_j with $j \neq 1$. Then each admissible tuple for $\{(x_1, I_1), \ldots, (x_k, I_k)\}$ can be derived from an admissible tuple of $(x_2, I_2), \ldots, (x_k, I_k)$, by rule 2.
3. Suppose I_1 does not cover all other I_i. Then the "undefined" vertex U has more than one child in the matching tree T, say $(x_{i_1}, I_{i_1}) \ldots, (x_{i_p}, I_{i_p})$.
   Let $D(x_l, I_l)$ be the set of descendents of (x_l, I_l) in the matching tree T. Then by proposition 3 the set $D(x_l, I_l)$ is of the form
   $$\{(x_{l+1}, I_{l+1}), \ldots, (x_{l+r}, I_{l+r})\}.$$
   Therefore M is the concatenation of $D(x_{i_1}, I_{i_1}) \cup \{(x_{i_1}, I_{i_1})\}, \ldots, D(x_{i_p}, I_{i_p}) \cup \{(x_{i_p}, I_{i_p})\}$, i.e.
   Each x_i appearing in some $D(x_{i_j}, I_{i_j}) \cup \{(x_{i_j}, I_{i_j})\}$, such that $j < p$ is smaller than each x_i appearing in $D(x_{i_p}, I_{i_p}) \cup \{(x_{i_p}, I_{i_p})\}$.
   Consider the matching $M' = \bigcup_{j<p} D(x_{i_j}, I_{i_j}) \cup \{(x_{i_j}, I_{i_j})\}$. Then $M = M' \cup D(x_{i_p}, I_{i_p}) \cup \{(x_{i_p}, I_{i_p})\}$. Moreover I_{i_p} is not contained in any I_j which appears in M'. Otherwise the parent of I_{i_p} would not be undefined. Therefore M can be generated from M' and $D(x_{i_p}, I_{i_p}) \cup \{(x_{i_p}, I_{i_p})\}$ using proposition 4 and 5. That means the admissible tuples for M arise from the admissible tuples of M' and of $D(x_{i_p}, I_{i_p}) \cup \{(x_{i_p}, I_{i_p})\}$ by the use of rule 1.

Theorem: *A maximum cardinality alternating cycle free matching and therefore the minimum jump number for convex bipartite graphs can be found in polynomial time.*

Proof: Since the number of admissible tuples is polynomially bounded, we get by iterative application of rules 1 and 2 admissible tuples in polynomial time. Therefore we get also the maximal cardinality of an alternating cycle free matching in polynomial time.

To get such a matching explicitly we construct a derivation tree T' as follows: We pick up an admissible tuple t associated to some alternating cycle free matching of maximal size. We also pick up a rule 1 or 2 and two or one admissible tuples t_1 and t_2 or t_1 respectively, from which t can be derived by one application of the corresponding rule. t_1 and t_2 or t_1 respectively are the children of t. For each leaf t', we generate children t'_1 (and t'_2) by this way, as long it cannot be derived by the initialization rule.

By a simple induction, we can prove that the set of all pairs (x, I) appearing in some leaf or some application of rule 2 in the tree T', is an alternating cycle free matching for which t is an admissible tuple. Therefore also a maximum alternating cycle free matching can be constructed in polynomial time.

□

3 Improvement of the Algorithm

Note that there are $O(n^8)$ admissible tuples and therefore $O(n^{16})$ possible applications of rule 1. This is a polynomial time bound but a bad polynomial time bound.

A closer look to the proof of proposition 9 shows that we can restrict the application of rule 1 to the case that the second matching $(\bar{x_1}, \bar{x_2}, \bar{u}, \bar{v}, \bar{v}_{min}, \bar{v}_{max}, \bar{r}, \bar{s}, \bar{m})$ corresponds to a matching such that the first interval covers all remaining intervals, i.e. $\bar{u} = \bar{v}_{min}$ and $\bar{v} = \bar{v}_{max}$. Therefore we make use of the first interval only in case that it covers all other intervals of the matching. Moreover, note that in each alternating cycle free matching $M = \{(x_j, I_j) : j = 1, \dots k\}$ with right most sequence $((x_{i_j}, I_{i_j}) : j = 1, \dots q)$, the largest index k belongs to the right most sequence $(k = i_q)$ and therefore x_k is less than the right border of any I_{i_j}. Therefore, in the notation of the right most sequence, $x_2^M = x_k < s_j^M$ for $j < q$. Therefore we have to check in rule 1 that $\bar{x_1} > x_2$ only in the case that s is an s_q^M, i.e. $s = -\infty$. Therefore it is reasonable to redefine $s_q^M = x_k = x_2^M$. In all other cases, we do not make use of x_2^M.

Therefore we need only to store x_1^M, v_{min}^M, v_{max}^M, r_j^M, s_j^M, $\#M$, and a flag f that is 2 if the first interval covers the remaining intervals and 1 otherwise.

The new rules to get *reduced admissible tuples* $(x_1^M, v_{min}^M, v_{max}^M, r_j^M, s_j^M, f, \#M)$ are the following:

Initialization If $I = [u, v] \in \mathcal{I}$ and $x \in I$ then

$$(x, u, v, x, x, 2, 1)$$

is reduced admissible.

Rule 1 Suppose $(x_1, v_{min}, v_{max}, r, s, f, m)$ and $(\bar{x}_1, \bar{v}_{min}, \bar{v}_{max}, \bar{r}, \bar{s}, 2, \bar{m})$ are reduced admissible and the following statements are satisfied:

$$v_{max} < \bar{v}_{max} \text{ and}$$

$$v_{max} < \bar{x}_1 \text{ or } (r < \bar{v}_{min} \, and \, s < \bar{x}_1).$$

Then

$$(x_1, \min(v_{min}, \bar{v}_m in), \bar{v}_{max}, \bar{r}, \bar{s}, 1, m + \bar{m})$$

is reduced admissible.

Rule 2 Suppose that $(x_1, v_{min}, v_{max}, r, s, f, m)$ is reduced admissible, $x \in I = [\tilde{u}, \tilde{v}] \in \mathcal{I}$, and that the following statements are true:

$$x < v_{min}, v_{max} < \tilde{v}.$$

Then

$$(x, \tilde{u}, \tilde{v}, r, s, 2, m + 1)$$

and

$$(x, \tilde{u}, \tilde{v}, x, v_{max}, 2, m + 1)$$

are reduced admissible.

The number of reduced admissible tuples $(x, u, v, r, s, 1, m)$ is bounded by n^6 and the number of reduced admissible tuples $(x, u, v, r, s, 2, m)$ is bounded n^5 (note that we have $\leq n$ intervals $[u, v]$).

Rule 1 can be applied efficiently as follows. For each reduced admissible

$$(x_1, v_{min}, v_{max}, r, s, f, m)$$

and each $\bar{v}_{min}$, $\bar{v}_{max}$, $\bar{x}_1$, such that rule 1 applies, we store a *preadmissible tuple*

$$(x_1, v_{min}, \bar{x}_1, \bar{v}_{min}, \bar{v}_{max}, m).$$

If

$$(x_1, v_{min}, v_{max}, \bar{x}_1, \bar{v}_{min}, \bar{v}_{max}, m)$$

is preadmissible and

$$(\bar{x}_1, \bar{v}_{min}, \bar{v}_{max}, \bar{r}, \bar{s}, 2, \bar{m})$$

is reduced admissible then we make

$$(x_1, \min(v_{min}, \bar{v}_{min}), \bar{v}_{max}, \bar{r}, \bar{s}, 2, m + \bar{m})$$

admissible. The time complexity to get all preadmissible tuples can be bounded by $O(n^8)$ where n is the number of vertices of the graph, and the time complexity to apply rule 2 with the help of preadmissible tuples can be bounded by $O(n^9)$.

The time complexity to apply rule 2 can be bounded by $O(n^8)$

Therefore we get an overall time bound of $O(n^9)$.

4 Final Remarks

Still a time bound of $O(n^9)$ is very high. *It remains an interesting problem to get a better polynomial time algorithm for the jump number problem of convex bipartite graphs.*

References

1. A. Brandstädt, *The Jump Number Problem for Biconvex Graphs and Rectangle Covers of Rectangular Regions*, Fundamentals of Computation Theory (J. Csirik, J. Demetrovics, F. Gecseg ed.), LNCS 380, 1989, pp. 68-77.
2. S. Cook, *A Taxonomy of Problems with Fast Parallel Algorithms*, Information and Control 64 (1985), pp. 2-22.
3. G. Chaty, M. Chein, *Ordered Matchings and Matchings without Alternating Cycles in Bipartite Graphs*, Utilitas Mathematica 16 (1979), pp. 183-187.
4. H. Müller, *Alternating Cycle Free Matchings in Chordal Bipartite Graphs*, Order 7 (1990), pp. 11-21.

Fast Lattice Browsing on Sparse Representation

Maurizio Talamo[1] and Paola Vocca[2]

[1] Dipartimento di Informatica e Sistemistica, Università di Roma "La Sapienza", Via Salaria 113, I-00198 Rome, Italy
[2] Dipartimento di Matematica, Università di Roma "Tor Vergata", Via della Ricerca Scientifica, I-00173 Rome, Italy.
E-mail vocca@mat.utovrm.it

Abstract.
In this paper, we present an implicit data structure for partial lattices representation, which allows to efficiently perform, with respect to either time and space, the following operations: 1) testing partial order relation among two given elements and 2) given the Hasse diagram representation and two related elements u and v, returning a sequence $< u_1, \ldots, u_l >$ of elements such that $u \prec u_1 \prec \ldots \prec u_l \prec v$. This first operation can be performed in constant time while the second in time $O(l)$, where l is the sequence size. The data structure proposed has an overall $O(n\sqrt{n})$-space complexity which we will prove to be optimal in the worst case. Hence, we derive an overall $O(n\sqrt{n})$ -space*time bound for the relation testing problem so beating the $O(n^2)$ bottle-neck representing the present complexity.
The overall pre-processing time is $O(n^2)$.

1 Introduction.

The study of partial orders efficient representation with respect both to time and space has been extensively tackled in the last years for many problems as recursive queries management in a DBMS [19, 5], dictionary problem [4, 6]; navigation within knowledge bases [1, 2], object-oriented and semantic data models [11], expert systems, VLSI layout, computational geometry [8], and distributed computing [14, 15]. In general, the partial order representation is involved in all that applications dealing with traversing sets of items over which an order relation is defined.

In this paper, we are mainly concerned in the representation of a *partial lattice* $\mathcal{L}(\prec, \mathcal{N})$, where a partial lattice is such that, adding to $\mathcal{N}$ two elements s and t, and extending $\prec$ with the following order relations: $s \prec x$ and $x \prec t$, $\forall x \in \mathcal{N}$, the resulting partial order is a lattice. Other authors refers to this class of posets as *truncated lattices* [16].

In particular, given a partial lattice $\mathcal{L}(\prec, \mathcal{N})$, with $n = |\mathcal{N}|$, we want to perform the following operations:

* Work partially supported by the ESPRIT Basic Research Action No.7141 (ALCOM II)

1. Given $u, v \in \mathcal{N}$, return true if $u \prec v$; false otherwise;
2. Given the Hasse Diagram representation of $\mathcal{L}$ and two related elements $u, v \in \mathcal{N}$, return a sequence $< u_1, \ldots, u_l >$ of elements such that $u \prec u_1 \prec \ldots \prec u_l \prec y$.

The target is to minimize the storage complexity while efficiently performing the above operations.

Even if a lot of work has been done in this field, it is still an open problem how to maintain partial order relations with a worst case time*space complexity less then the usual ϵn^2 (ϵ a positive constant). Therefore, only for restricted classes of partial orders an efficient solution to the above problem has been found [10, 15, 17] exploiting the order dimension property of posets [7]. In this context, the posets considered are those having constant order dimension. Unfortunately, not for all posets the order dimension is constant, in general, it is function of the size of the element set size. Hence, this technique can not be extended to deal with general posets. In particular, this is true for partial lattices. The recently introduced *boolean dimension* of partial orders [9], a generalization of the classical order dimension, seems to represent a promising approach for overcoming the above problem.

Anyway, in this paper, the approach followed is different; we adopt a two-level decomposition strategy of the associated Hasse Diagram which allows to exploit specific structural properties of the class of posets under investigation.

In fact, a general problem of a decomposition technique is the choice of the decomposition criteria as the more it takes into account space complexity, the more query time is expensive, and viceversa. The basic idea underlying our technique for balancing space complexity versus time complexity, is to adopt a double decomposition criteria, the first for enhancing the representation space, while the second for maintaining query efficiency.

With this strategy we derive a sparse representation allowing to perform a constant partial order relation test having a worst case $O(n\sqrt{n})$-space complexity which is *sublinear* in the number of ordered pairs necessary to encode arbitrary partial orders.

More formally, the following theorem will be proved:

Theorem 1. *Let $\mathcal{L}(\prec, \mathcal{N})$ a partial lattice, where $n = |\mathcal{N}|$, a $O(n\sqrt{n})$ -space* implicit *data structure exists, allowing to test partial order relation in $O(1)$-time and to retrieve a total order on the Hasse Diagram $O(l)$-time, where l is the size of the returned total order. The overall pre-processing time is $O(n^2)$.*

The bound obtained not only does not break any information theoretic lower bound, but is optimal with respect to the worst case as shown by the following proposition:

Proposition 2 [13]. *Let $L(n)$ the number of labelled lattices with n elements then $L(n) < \alpha^{(n\sqrt{n}+o(n\sqrt{n}))}$, where α is a constant (about 6.11343).*

Results proved in this paper have a natural field of application to the transitive closure of digraph representation problem. In fact, given a directed acyclic graph $G = (N, A)$, let $\mathcal{P}_G(\prec_\mathcal{P}, N)$ the *associated partial order*, that is, $u \prec_P v$, with $u, v \in N$, if and only if $< u, v > \in A^*$, where A^* is the edge set of the transitive closure graph $G^* = (N, A^*)$. In particular, we are interested in representing graphs whose associated partial order is a partial lattice. We denote this class of graph *digraph satisfying the lattice property.*

Using a graph-theoretic terminology, theorem 1 can be formulated as follows:

Theorem 3. *Let $G = (N, A)$ a directed acyclic graph (dag) satisfying lattice property, with $n = |N|$. An $O(n\sqrt{n})$-space implicit data structure exists allowing to perform the following operations:*

a. REACHABILITY(u, v)*: Given $u, v \in N$, test the presence of a directed path from u to v: return true if such a path exists; false otherwise;*
b. PATH(u,v)*: given $u, v \in N$, it returns an ordered sequence $< x_0, x_1, \ldots, x_l >$ of vertices, such that $x_0 = u$, $x_l = v$, and $< x_i, x_{i+1} > \in A$ for $i = 0, \ldots, l-1$, if at least one path exists; undefined otherwise.*

The first operations can be performed in $O(1)$-time, while the second in $O(l)$. The overall pre-processing time is $O(n^2)$.

The paper is organized as follows: section 2, is mainly concerned with the description of the basic decomposition structures; in section 3, the decomposition strategy is described and general space and time bounds are given; in section 4, the concept of decomposition *base* is introduced and its main properties are stated. As described in section 5, the base definition allows to derive to a new version of the decomposition strategy leading to the desired complexity bounds. Finally, in section 6, applications and future researches are described.

2 Preliminaries.

In this section, we describe an important structural property of the class of partial orders under investigation and introduce the basic decomposition structures showing their properties.

Proposition 4. *A partial order $\mathcal{L}(\prec, \mathcal{N})$ is a partial lattice if and only if for every four elements $(u, v, z, w) \in \mathcal{N}$, if four directed paths in $HD(\mathcal{L})$ exist, pairwise disjoint except for at most the extremal elements, having as endpoints the following couple of elements $< u, z >$, $< u, w >$, $< v, z >$, and $< v, w >$, then an element x and four paths having as endpoints the following couples nodes $< u, x >$, $< v, x >$, $< x, z >$, and $< x, w >$ do exist.*

Figure 1 shows the above proposition. The obvious interpretation is the following: two elements u and v can't have two different greatest lower bounds, w and z, without the existence of a fifth vertex x representing the actual greatest lower bound. Analogously, two vertices w and z can't have two different least upper bounds, u and v, without the existence of a fifth vertex x representing the actual least upper bound.

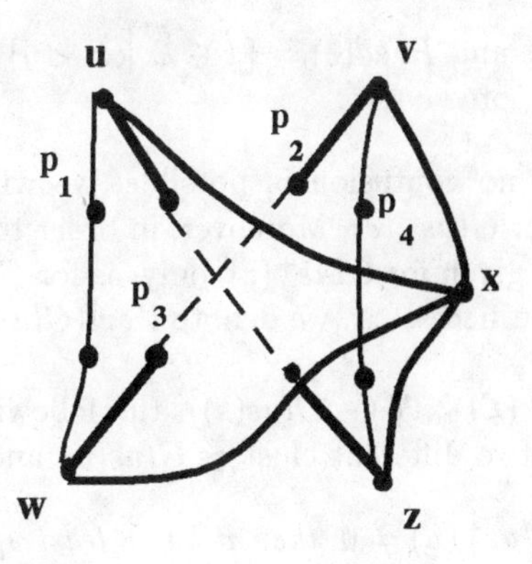

Fig. 1. Partial Lattices Property.

We will use the above property for granting the correcteness of our approach. In fact, as it will be evident in the following, it allows to obtain a constant time partial order relation test.

2.1 Cluster

Let us now introduce the basic decomposition structures of the double decomposition strategy proposed.

Given a partial lattice $\mathcal{L}(\prec, \mathcal{N})$, on the first level we partition it in a sequence of forests of disjoint partial orders denoted *cluster* or $Clus(c)$, where $c \in \mathcal{N}$.

On the second level, for each cluster $Clus(c)$, we choose a suitable collection of posets (*double-tree* or $DT(u, c)$) representing all partial order relations between elements in $Clus(c)$ and elements in $(\mathcal{L}(\prec, \mathcal{N}) - Clus(c))$.

More precisely, for every two elements x, y, such that $x \in Clus(c)$, the decomposition in $\{DT(u, c)\}$ satisfies the following property: $x \prec y$ in $\mathcal{L}(\prec, \mathcal{N})$ if and only if it exists at least one $DT(u_i, c) \in \{DT(u, c)\}$ such that $x \prec y$ in $DT(u_i, c)$.

As the first level decomposition is a partition of the element set, we will show that, operating on the cluster size, it is possible to bound the overall space complexity, while granting, with the second level decomposition, query consistency.

Nevertheless, we first illustrate the decomposition technique and the corresponding data structure regardeless whichever the cluster is, and then, by means of the decomposition base definition (see section 4), we will prove the required space bounds.

We now give a formal definition of cluster and state its main properties:

Definition 5. Given an element $c \in \mathcal{N}$, let $Clus^+(c)$ and $Clus^-(c)$ be the posets induced by the sets $\{Succ(c) \cup c\}$ and $\{Pred(c) \cup c\}$, respectively, where

$Succ(c) = \{x \in \mathcal{L}, ; |; c \prec x\}$ and $Pred(c) = \{x \in \mathcal{L}; |; x \prec y\}$. A *cluster* is either $Clus^+(c)$ or $Clus^-(c)$ for some $c \in \mathcal{N}$.

In the following, when no confusion is possible, we will use $Clus(c)$ for denoting either $Clus^+(c)$ or $Clus^-(c)$. Moreover, in order to make proofs more readable, all our results are given for $Clus^+(c)$ only, as for $Clus^-(c)$ proofs can be dually derived. Yet, when necessary, we denote $Vert(Clus(c))$ the ground set of $Clus(c)$.

Let u be an element in $(\mathcal{L}(\prec, \mathcal{N}) - Clus(c))$, the following lemma shows a basic relationship between two different clusters $Clus(c)$ and $Clus(u)$.

Lemma 6. *If* $Clus^+(c) \cap Clus^+(u) \neq \emptyset$ *then it has a least upper bound. Dually, if* $Clus^-(c) \cap Clus^-(u) \neq \emptyset$ *then it has a greatest lower bound.*

Proof. By cluster definition, it follows that for every two elements $x, y \in Clus^+(c)$, $LUB(x, y) \in Clus^+(c)$. In fact, assume, for a contradiction, that $LUB(x, y) \notin Clus^+(c)$. Then the following three conditions simultaneously hold:

i. $x \prec c$
ii. $y \prec c$
iii. $LUB(x, y) \sim c$

This contradicts proposition 4.

In order to prove the lemma we have to show that $Clus^+(c) \cap Clus^+(u)$ doesn't have two distinct maximal elements. Let us now assume that two maximal elements $x, y \in Clus^+(c) \cap Clus^+(u)$ exist, with $x \neq y$. Then we have $LUB(x, y) \sim u$, but this, once again contradicts proposition 4. The lemma is so proved by contradiction. □

Lemma 7. *Let* $\mathcal{L}(\prec, \mathcal{N})$ *a partial lattice and* $Clus(c)$ *a cluster, then* $(\mathcal{L}(\prec, \mathcal{N}) - Clus(c))$ *is a partial lattice.*

Proof. It is an obvious consequence of cluster definition. □

2.2 Double-Tree

Given a cluster $Clus(c)$, let us refer to all elements of $Clus(c)$ and in $(\mathcal{L}(\prec, \mathcal{N}) - Clus(c))$ as *internal* and *external elements*, respectively. In particular, we denote $Int(Clus(c))$ the set of internal elements and $Ext(Clus(c))$ the set of all external elements related to at least one internal element.

In order to efficiently perform the order relation test, a suitable representation for all connectivity information related to a given cluster $Clus(c)$ is needed. In particular, we have to face two main problems: *i)* the representation of the partial order relations between internal elements; *ii)* the representation of the partial order relations between internal and external elements.

The first problem can be easily solved computing for each internal element u the spanning tree, rooted at u, of the Hasse Diagram representing the set of all internal elements related to u. We refer to this tree as *Internal Tree induced by* u, denoted $IntTree(u, c)$.

For what regards the second problem, from lemma 6 it derives that, given an external vertex v, the couple $(v, Clus(c))$, univocally identifies a vertex $u \in Clus(c)$, representing, either the LUB($Clus^+(c) \cap Clus^+(v)$) or the GLB($Clus^-(c) \cap Clus^-(v)$) .

This implies that, given a cluster $Clus(c)$, for each external element v, related to at least one element in $Clus(c)$, an internal element u, the *internal representative of the external element* v exists, univocally identifying the internal tree $IntTree(u, c)$ made up with all elements in $Clus(c)$ related to v.

Let $Ext(u)$ be the set of external elements having u as internal representative. Obviously, it follows:

$$Ext(Clus(c)) = \bigcup_{u \in \{Clus(c)\}} Ext(u)$$

For each set in the collection $\{Ext(u)\}$ we compute a spanning tree of the associated Hasse Diagram, starting from the internal representative u. We refer to this tree as *External Tree induced by* u, denoted $ExtTree(u, c)$ (see figure 2). Moreover, the collection of external trees $\{ExtTree(u_i, c)\}$ has the nice property to be pairwise disjoint, as shown by the following lemma:

Lemma 8. *Let* $v, w \in Clus(c)$, *with* $v \neq w$ *then* $ExtTree(v, c) \cap ExtTree(w, c) = \emptyset$.

Proof. Let us assume, for a contradiction, $y \in ExtTree(v, c) \cap ExtTree(w, c)$. By external tree definition, it derives that y is associated to both v and w. This contradicts the uniqueness of representative element stated in lemma 6. The proof follows by contradiction. □

Definition 9. Given a cluster $Clus(c)$ and the two collections of internal and external trees, $\{IntTree(u_i, c)\}$ and $\{ExtTree(u_i, c)\}$, for each $u_i \in Clus(c)$ a *Double Tree* is defined as

$$DT(u_i, c) = IntTree(u_i, c) \cup ExtTree(u_i, c).$$

A double tree represents the second level decomposition structure, and informally speaking, is the union of an internal tree and an external tree rooted at the same internal element, whenever they exist.

Each double tree, considered as the union of two rooted trees, is associated to a partial order having order dimension 2, as its st-completion is a planar poset with one greatest element (source) and one least element (sink) [18]. The first consequence of the above property is that it is possible to find two linear extensions L_1, L_2 representing the partial order $DT(u, c)$, that is given to vertices $x, y \in DT(u, c)$, $x \prec y$ if and only if $x \prec y$ in both linear extensions L_1, L_2. In particular, two labels (*coordinates*) (x_1, x_2) are associated to each vertex x, each one representing x position within the first and the second linear extension, respectively. The following proposition holds and its proof can be found in [12]:

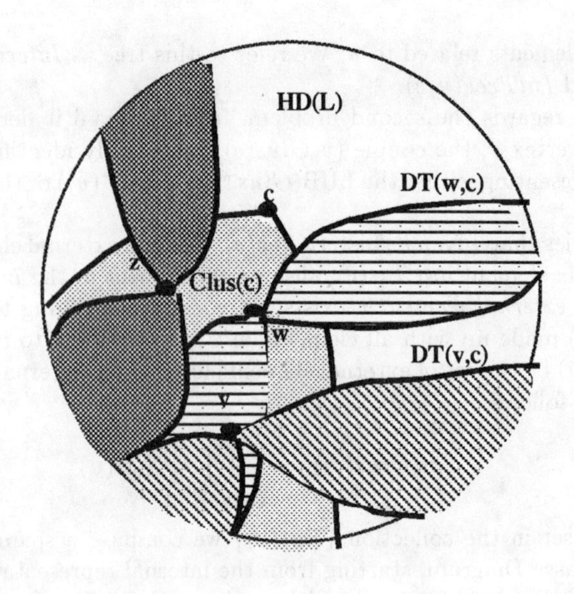

Fig. 2. Double-Trees Decomposition.

Proposition 10. *Given $x, y \in DT(u,c)$, $x \prec y$ in $DT(u,c)$ if and only if $(x_1, x_2) < (y_1, y_2)$.*

From the above proposition it easily follows:

Corollary 11. *Given a double tree $DT(u,c)$, such that $|DT(u,c)| = N$, it exists a $O(N)$-space data structure for testing partial order relation among two elements in $O(1)$-time.*

3 Decomposition Strategy and Data Structure.

Before formally describing our decomposition strategy, we have to give two further definitions.

Definition 12. Two clusters $Clus(c_1)$ and $Clus(c_2)$, where $c_1, c_2 \in \mathcal{N}$, are *independent* if and only if $Clus(c_1) \cap Clus(c_2) = \emptyset$.

Definition 13. A set of clusters $F = \{Clus(c_1), \ldots, Clus(c_k)\}$ is a *cluster forest* if all clusters are pairwise independent.

The decomposition strategy we propose builds at each main iteration a forest of clusters, and for each cluster builds the associated double-trees structure.

procedure DECOMPOSITION1 $(\mathcal{L}(\prec, \mathcal{N}), \mathcal{F} = \{F_i\}, \mathcal{T} = \{DT(u, c_{ij})\})$;

```
1. begin;
2. 𝓕 := {∅};          {Collection of Clusters Forests}
3. 𝓣 := {∅};          {Collection of Double-Trees}
4.   while 𝓝 ≠ {∅} do
5.     F := {∅};
6.     while at least one cluster indipendent with F exists do
7.        Choose the largest cluster Clus(c) indipendent with F;
          {either Clus⁻(c) or Clus⁺(c)}
8.        F := F ∪ Clus(c);
9.        for each u ∈ Clus(c) do
10.          Build DT(u, c);
11.          𝓣 = 𝓣 ∪ DTree(u, c);
12.       end;
13.       𝓝 := 𝓝 − Vert(Clus(c));
14.    end;
15.    𝓕 := 𝓕 ∪ F
16.end
17.end.
```

DECOMPOSITION1 procedure generates two collections of subgraphs as follows:

$\mathcal{F} = \{F_1, \ldots, F_h\}$ where $F_i = \{Clus(c_{i1}), \ldots, Clus(c_{ik_i})\}$

$\mathcal{T} = \{DT(u, c_{i,j}),\}$ where $i = 1, \ldots, h$ and $j = 1, \ldots, k_i$

Collection $\mathcal{F}$ is an ordered set of h cluster forests, while $\mathcal{T}$ is made up of $\sum_{i=1}^{h} \sum_{j=1}^{k_i} t_{ij}$ double-trees, where $t_{ij} = |Vert(Clus(c_{ij}))|$.

Theorem 14 proves the correctness of the above algorithm showing that the decomposition obtained is a covering of the given partial lattice. In fact, the strategy proposed is quite general and it can be applied to partial lattices obtaining, as it shown in the following, the required time bounds. Neverthless, in order to derive the desired space bounds, additional conditions are needed (see DECOMPOSITION2 procedure of section 5.

Theorem 14. *Given a partial lattice $\mathcal{L}(\prec, \mathcal{N})$ and $x, y \in \mathcal{N}$, $x \prec y$ in $\mathcal{L}$ if and only if $x \prec y$ in $DT(v, c)$, for at least one $DT(v, c) \in \mathcal{T}$.*

Proof. ($\Rightarrow$) By lemma 7, at each main iteration, the partial order $(\mathcal{L}(\prec, \mathcal{N}) - Clus(c))$ is still a partial lattice. Let $x \prec y$ in $\mathcal{L}$. We have to show that the DECOMPOSITION1 procedure returns a $DT(u, c) \in \mathcal{T}$ such that $x \prec y$ in $DT(u, c)$. Note that, by the termination condition (line 4), x and y must belong to one and only one cluster. Two different cases are possible according to which cluster x and y belong to.

a. $x, y \in Clus(c) \in F_i$.
   If $Clus(c)$ is $Clus^+(c)$, then lines 9- 12 assure that y belongs to $IntTree(x, c)$, hence $y \in DT(x, c)$. On the other hand, if $Clus(c)$ is a $Clus^-(c)$ then $x \in DT(y, c)$.

b. $x \in Clus(c_1)$ and $y \in Clus(c_2)$.
Let's suppose that the DECOMPOSITION1 procedure first generates $Clus(c_1)$ and then $Clus(c_2)$. By double-tree definition, y belongs to $ExtTree(x, c_1)$, hence $y \in DT(x, c_1)$. Analogously, if $Clus(c_2)$ has been generated before $Clus(c_1)$ then $x \in DT(y, c_2)$.

($\Leftarrow$) This part of the proof obviously follows observing that the decomposition algorithm doesn't add any new partial order relation. □

Given the two above collections and an element $x \in \mathcal{N}$, the following assertions hold:

i. x belongs to one and only one forest F_i; (for construction, line 13);
ii. x belongs to one and only one cluster $Clus(c_{ij}) \in F_i$;
iii. given a forest F_j, different from the one to which x belongs to according *i*), and a cluster $Clus(c_{jl}) \in F_j$, then x belongs to at most one $DT(u, c_{jl})$ (lemma 8);
iv. given a cluster $Clus(c)$, each element of the collection $\{DT(u_i, c)\}$ is univocally identified by the element $u_i \in Clus(c)$ (lemma 6);
v. if $Clus(c)$ a $Clus^+(c)$, let $DT(u, c)$ an associated double tree. If $x \in IntTree(u, c)$ with $x \neq u$, then x has at most one predecessor in $DT(u, c)$. Dually, if $x \in ExtTree(u, c)$, then x has at most one successor in $DT(u, c)$. If $Clus(c)$ is a $Clus^-(c)$, the argumentation is similar.

From the above observations and applying proposition 10 for double-trees representation, we derive the following implicit data structure.

The data structure is based on one look-up table indexed on vertices in $\mathcal{N}$(data structure $\mathcal{A}$), and three sets of look-up tables (data structures $\mathcal{B}_x$, $\mathcal{C}_x$, and $\mathcal{D}_x(F_i)$).

Data structure $\mathcal{A}$ stores for each vertex the unique identifier of forest F_i and cluster $Clus(c_{ij})$ at which it belongs to (oss. *i.* and *ii.*).

Data structure $\mathcal{B}$, is a set of look up-table each one associated to an element $x\mathcal{N}$. For each double tree $DT(u, c_{ij})$ of the decomposition induced by the cluster $Clus(c_{ij})$ to which x belongs, data structure $\mathcal{B}_x$ stores x and x's parent coordinates with respect to the representation of the double tree $DT(u, c_{ij})$, whenever x belongs to $DTree(u, c_{ij})$; otherwise it contains a null value.

Also, data structure $\mathcal{C}$, is a set of look up-table each one associated to an element $x\mathcal{N}$. For each forest F_i of the decomposition, if x is related to at least one cluster in F_i, i.e. $x \in Ext(F_i)$, data structure $\mathcal{C}_x$ stores the identifier of a $\mathcal{D}$ table storing all partial order relations between x and F_i. If x is not related to any element in F_i then it stores a null value.

Finally, data structure $\mathcal{D}$, is a set of look up-table each one associated to an element x and a forests F_i. Table $\mathcal{D}_x(F_i)$ exists only if $x \in Ext(F_i)$. For each cluster $Clus(c)$, componing the forest F_i, the corresponding field of look-up table $\mathcal{D}_x(F_i)$ store the identifier of the double tree associated to $Clus(c)$, at which x belongs to as external element. Moreover, the field contains x and x's parent coordinates.

In order to state general space bounds, it is useful to introduce the following notations:

- $|\mathcal{F}| = h$
- $|Vert(Clus(c_{ij}))| = t_{ij}$
- $t = Max\{t_{ij}\}$
- $m_i = |Ext(F_i)|$.

Lemma 15. *Let $\mathcal{L}(\prec, \mathcal{N})$ a partial lattice, an $O(nt + nh + \sum_{i=1}^{h} m_i k_i)$-space data structure exists allowing to test the partial order relation between two given elements in $O(1)$-time and to return the sequence $< x_1, \ldots, x_l >$ of elements, such that $x \prec x_1 \prec \ldots \prec x_l \prec y$ in $HD(\mathcal{L})$, in $O(l)$-time, l is the sequence lenght.*

Proof. The first part of the lemma follows observing that:

i. Data structure $\mathcal{A}$: $O(n)$;
ii. Data structure $\mathcal{B}$: $O(nt)$. Each look-up table $\mathcal{B}_x$ is $O(t_{ij})$. In fact, if $t_{ij} = |Vert(Clus(c_{ij}))|$, then at most t_{ij} different double-trees are possible, one for each element in $Clus(c_{ij})$.
iii. Data structure $\mathcal{C}$: $O(nh)$
iv. Data structure $\mathcal{D}$: $O(\sum_{i=1}^{h} m_i k_i)$ as only vertices y connected to a forest have the corresponding $\mathcal{D}_y$ look-up table, each one of size $O(k_i)$.

For what regards the partial order relation testing, by theorem 14, the central problem is to find the double tree $DT(u, c)$ at which both x and y belong to. This can be done by first looking in the data structure $\mathcal{A}$ for forest and cluster identifiers. If x and y belong to the same forest, then we search the look-up tables $\mathcal{B}_x$ and $\mathcal{B}_y$ for the double-tree coordinates of x and y with respect to: i) the double tree rooted at x; ii) the double tree rooted at y. If in one of the two double-trees, y's coordinates are greater than x's ones then $x \prec y$, otherwise not.

If x and y doesn't belong to the same forest then let $x \in F_i$ and $y \in F_j$. We first search in the table $\mathcal{D}_y(F_i)$ the identifier of the double tree associated to the cluster at which x belongs to. Then we search in the $\mathcal{B}_x$ table for the x coordinates with respect to this double tree. If y coordinates are greater than x ones then $x \prec y$. If the above search fails, we can repeat the same strategy, looking in the $\mathcal{D}_x(F_j)$ for the identifier of the double tree associated to the cluster at which y belongs to. Then we search in the $\mathcal{B}_y$ table for the y coordinates with respect to this double tree. If y coordinates are greater then x ones then $x \prec y$. If both these searches fail, then $x \not\prec y$. Obviously, the above strategy can be implemented in $O(1)$.

For what concerns the path operation on the Hasse Diagram, once tested if $x \prec y$ and found the double tree at which they both belong to, we search the father coordinates of both x and y (data structure $\mathcal{B}$ and $\mathcal{D}$). The search goes on until these coordinates are equal. For this operation it is important the kind of cluster at which x and y belong to, as this allows derive the search direction. □

In the following sections, the overall space complexity is bounded introducing an *ad hoc* forest decomposition

4 Decomposition Base Definition.

In this section we show how to bound space complexity by a proper choice of clusters and forests of clusters. We claim that an overall $O(\sqrt{n})$-size for forests is in fact the answer to our problem as this size allows to balance what it is eliminated in one main iteration and what remains to consider. In fact, at each main iteration of the DECOMPOSITION1 procedure (line 4), we eliminate, from future analysis, the whole partial order induced by F_i, and, as a consequence, we have to be sure that its size is large enough to guarantee an overall efficient decomposition.

In order to formalize the above strategy, we introduce the definition of *decomposition base.* Informally speaking, a decomposition base is a suitable element inducing a cluster with the required size.

Definition 16. Given $\mathcal{L}(\prec, \mathcal{N})$, a vertex $c \in \mathcal{N}$, such that $|Vert(Clus(c))| = a\sqrt{n}$ is a *base* for $\mathcal{L}$ if one of the following two conditions holds:

$$\textit{for } a < \tfrac{1}{4} \ \sum_{u \in Clus(c)} |ExtTree(u, c)| \leq \tfrac{n}{16}$$

or

$$\textit{for } a \geq \tfrac{1}{4} \ \sum_{u \in Clus(c)} |IntTree(u, c)| \leq \tfrac{n}{16}$$

This definition allows us to evaluate the complexity of what must be represented at each iteration of the DECOMPOSITION1 algorithm, i.e. all elements related to the cluster under consideration. The first part of the definition identifies bases generating clusters of smaller size. These clusters can then be grouped together to form a forest. The second part identifies bases generating clusters of the required size.

The following theorem ensures that it always exists a cluster decomposition induced by bases. In particular, we have to show that given a partial lattice it is always possible to find a base.

Theorem 17. *A partial lattice $\mathcal{L}(\prec, \mathcal{N})$ has got at least one base which can be found in time $O(n^2)$.*

Proof. Let us denote $U, L \subseteq \mathcal{N}$, the sets of maximals and minimals elements in $\mathcal{N}$. Without loss of generality, let us consider a layers embedding of the Hasse Diagram $HD(\mathcal{L})$. The lower level represents L, while the set U correspond to the upper level.

We first look for a base satisfying the second condition. If this search fails then a base satisfying the first condition is considered.

In particular, we first search for a base inducing a $Clus^{+}(c)$, starting from an element in L, and follows backward a path to an element in U. Then, if the above strategy fails, a dual search starts from an element in U for a base inducing a $Clus^{-}(c)$.

The search can lead to five different cases which are described in the following.

a. If a father y of a minimal element x has outdegree greater or equal to $\frac{\sqrt{n}}{4}$, then y is a base. In fact we have:

$$\sum_{u \in Clus^+(y)} |IntTree(u, y)| \leq \frac{n}{16}$$

as none of y's successors can be a base.

b. If on the path backward to a maximal element we find an element y such that:

$$|Vert(Clus^+(y))| = \frac{\sqrt{n}}{4}$$

then y is a base. In fact, using an argumentation similar to case a) above, y satisfies base condition.

c. If a vertex y is met such that:

$$|Vert(Clus^+(y))| > \frac{\sqrt{n}}{4},$$

searching its successors two sub-cases, $c1$ and $c2$, are possible.

c1. In this case, a vertex $z \in Succ(y)$ exists such that:

$$|Vert(Clus^+(z))| = \frac{\sqrt{n}}{4}.$$

Using the same argumentation as above we can prove that z is base.

c2. Let us suppose that none of y successors satisfies the condition

$$|Vert(Clus^+(z))| = \frac{\sqrt{n}}{4},$$

then we recursively search the subgraph induced by y successors, for the cluster of smaller size satisfying the condition $|Vert(Clus^+(z))| > \frac{\sqrt{n}}{4}$. As it can be easily proved, z is a base.

f. If both searches for $Clus^+(c)$ and $Clus^-(c)$ satisfying the second condition fail, then the largest $Clus(c)$ is maintained and it is chosen as a base. Let t be the size of the maximal $Clus(c)$, i.e. $|Vert(Clus)| = t = a\sqrt{n}$ with $a < \frac{1}{4}$, then we have to show that $Clus(c)$ satisfies the first part of base definition. As $Clus(c)$has the largest size each $ExtTree(u, c)$ can't be greater than t. Observing that $Clus(c)$has t elements, the aim is proved as follows:

$$\sum_{u \in Clus(c)} |ExtTree(u, c)| \leq t^2 < \frac{n}{16}.$$

For what regards the time complexity, given the Hasse Diagram $HD(\mathcal{L})$, let us consider a data structure for representing it which maintains for each element the number of its successors and an ordered list of the number of successors of its immediate successors. This data structure can be derived in time $O(n^2)$ by recursively visiting $HD(\mathcal{L})$. With this data structure an algorithm for finding a base can be implemented in time linear in the number of elements. □

The above theorem assures the correctness of DECOMPOSITION2 procedure given in the following section and allows to evaluate time complexity of the preprocessing needed for constructing the proposed data structure.

5 Second Decomposition Algorithm and Space Complexity

In this section, a new version of the decomposition algorithm is given which bounds the overall space complexity by means of decomposition base definition.

Recall the definition of m_i as the size of $Ext(F_i)$, i.e. the number of elements belonging to $(\mathcal{L}(\prec,\mathcal{N}) - F_i)$ related to F_i, and of k_i as the number of independent clusters of F_i.

procedure DECOMPOSITION2 $(\mathcal{L}(\prec,\mathcal{N}), \mathcal{F} = \{F_i\}, \mathcal{T} = \{DT(u, c_{ij})\})$;

1. **begin;**
2. $\mathcal{F} := \{\emptyset\}$; {Collection of Forests of Clusters}
3. $\mathcal{T} := \{\emptyset\}$; {Collection of Double-Trees}
4. **while** $\mathcal{N} \neq \{\emptyset\}$ **do**
5. $F := \{\emptyset\}$;
6. $k := 0$; {Number of Indipendent Clusters in a Forest}
7. $m := 0$; {Number of External Elements Connected to a Forest}
8. **while at least one cluster indipendent with** F **exists and** $(|Vert(F)| < \frac{\sqrt{n}}{4})$ **and** $(mk < n)$ **do**
9. Choose a *base* $c \in \mathcal{N}$ s.t. $Clus(c)$ is the largest cluster independent with F; {either $Clus^-(c)$ or $Clus^+(c)$}
10. $F := F \cup Clus(c)$;
11. **for each** $u \in Clus(c)$ **do**
12. Build $DT(u, c)$;
13. $\mathcal{T} = \mathcal{T} \cup DT(u, v)$;
14. **end;**
15. $\mathcal{N} := \mathcal{N} - Vert(Clus(c))$;
16. $k := k + 1$;
17. $m := |Ext(F)|$;
18. **end;**
19. $\mathcal{F} := \mathcal{F} \cup F$;
20. **end;**
21. **end.**

We added to this version of the decomposition algorithm two new condition (see line 8) for bounding the overall space complexity.

In particular the i-th iteration of inner cycle of the above procedure ends when one of the following three conditions holds:

1. there are no more clusters independent with F_i;
2. $|Vert(F_i)| \geq \frac{\sqrt{n}}{4}$;
3. $m_i k_i \geq n$.

In order to prove the main theorem, we have to show that DECOMPOSITION2 procedure returns an $O(\sqrt{n})$ collection $\mathcal{F} = \{F_i\}$ of forests of clusters, i.e. at each iteration of the inner cycle at least $O(\sqrt{n})$ elements are taken.

Obviously, if the internal loop always terminates for condition 2 then the claim is proved. As a consequence, we have to analyse the other two cases.

Let us, first, introduce some more notation. Without loss of generality, we denote the size of a cluster $Clus(c_{ij}) \in F_i$:

$$|Vert(Clus(c_{ij}))| = n^{\frac{1}{2}-\sum_{p=1}^{j}\delta_{ip}} \tag{1}$$

where $\delta_{ip} \geq 0$, $\forall\, p \in \{1,\ldots,j\}$.

In fact, if the ordered sequence of clusters $< Clus(c_{i1}),\ldots,Clus(c_{ik_i}) >$ componing a forest, is generated from DECOMPOSITION2 procedure, then the corresponding sequence of cluster sizes is a monotone and not increasing, and by termination condition, each size is less than $\frac{\sqrt{n}}{4}$. Hence,

$$|Vert(F_i)| = n^{\frac{1}{2}}\sum_{j=1}^{k_i} n^{-\sum_{p=1}^{j}\delta_{ip}} \tag{2}$$

where the sequence $< \delta_{i1},\ldots,\delta_{ik_i} >$ is monotone not decreasing.

The following lemma shows that condition 1 never holds. In fact,

Lemma 18. *If* DECOMPOSITION2 *procedure ends for condition* 1 *then* $|Vert(F_i)| \geq \frac{\sqrt{n}}{4}$

Proof. Given a base $Clus(c_{ij})$ such that $|Vert(Clus(c_{ij}))| = t$, in order to find a base $Clus(c_{i,j+1})$ independent with $Clus(c_{i,j})$, elements in $Ext(Clus(c_{ij})$ cannot be considered. By base definition, the number of these elements is at most t^2. For hypothesis, the inner cycle of the DECOMPOSITION2 ends when no more clusters independent with F_i exist, that is, when the following condition holds:

$$\sum_{j=1}^{k_i}\left(n^{\frac{1}{2}-\sum_{p=1}^{j}\delta_{ip}}\right)^2 = \sum_{j=1}^{k_i} n^{1-2\sum_{p=1}^{j}\delta_{ip}} = n \implies \sum_{j=1}^{k_i} n^{-2\sum_{p=1}^{j}\delta_{ip}} = 1 \tag{3}$$

As the sequence $< n^{-\delta_{i1}}, n^{-\delta_{i1}-\delta i2},\ldots, n^{-\sum_{p=1}^{j}\delta_{ip}} >$ is a monotone not increasing sequence and each $n^{-\delta_{i1}} \leq 1$, from 3 follows:

$$\sum_{j=1}^{k_i} n^{-\sum_{p=1}^{j}\delta_{ip}} \geq 1$$

The claim is so proved. □

Let us, now analyse the case the DECOMPOSITION2 procedure ends for condition 3. Before, we prove the following lemma showing that forests with one cluster do not constitute a problem for our strategy.

Lemma 19. *If $k_i = 1$ then $m_i k_i = O(n)$.*

Proof. In fact, by base definition and termination condition:

$$m_i \leq |Vert(Clus(c_{i1})|^2 \leq \frac{n}{16}.$$

□

Let $\mathcal{F} =< F_1, \ldots, F_h >$, the sequence of cluster forests returned by the DECOMPOSITION2 procedure. Let F_{low}, where $1 \leq low \leq h$, the first forest with $k_{low} > 1$. Obviously, for $low \leq i \leq h$, $k_i > 1$.

If the i-th iteration, with $low \leq i \leq h$, ends for condition 3, i.e. $m_i k_i \geq n$, let $Clus(c_{i,k_i})$ the last cluster choosen. Then we have:

Lemma 20. $|Vert(Clus(c_{i,k_i}))| < \frac{n^{\frac{1}{2}\cdot \delta_{i1}}}{4}$.

Proof. By base definition, if $|Vert(Clus(c_{ij}))| = t_{ij}$ then the number m_{ij} of vertices connected to $Clus(c_{ij})$ is at most t_{ij}^2. As a consequence, we have:

$$m_i k_i \leq k_i \sum_{j=1}^{k_i} (t_{ij})^2 = k_i \sum_{j=1}^{k_i} \left(n^{\frac{1}{2} - \sum_{p=1}^{j} \delta_{ip}} \right)^2 = \tag{4}$$

$$k_i \sum_{j=1}^{k_i} n^{1-2\sum_{p=1}^{j} \delta_{ip}} \leq k_i \sum_{j=1}^{k_i} n^{1-2\delta_{i1}} = k_i^2 n^{1-2\delta_{i1}}. \tag{5}$$

Hence, by condition $m_i k_i \geq n$, it derives:

$$k_i^2 n^{-2\delta_{i1}} \geq 1 \Longrightarrow k_i \geq n^{\delta_{i1}} > 4 \tag{6}$$

The last inequalities derives from observing that if the algorithm terminates for condition 3, and not for condition 2 then:

$$|Vert(F_i)| = \sum_{j=1}^{k_i} n^{\frac{1}{2} - \sum_{p=1}^{j} \delta_{ip}} < \frac{\sqrt{n}}{4};$$

hence, $n^{-\delta_{i1}} < \frac{1}{4}$.
Moreover,

$$\frac{\sqrt{n}}{4} > \sum_{j=1}^{k_i} n^{\left(\frac{1}{2} - \sum_{p=1}^{j} \delta_{ip}\right)} \geq \sum_{j=1}^{k_i} n^{\left(\frac{1}{2} - \sum_{p=1}^{k_i} \delta_{ip}\right)} = k_i (n^{\left(\frac{1}{2} - \sum_{p=1}^{k_i} \delta_{ip}\right)}) \tag{7}$$

and, from relation 6 above:

$$k_i (n^{\left(\frac{1}{2} - \sum_{p=1}^{k_i} \delta_{ip}\right)}) \geq n^{\delta_{i1}} n^{\left(\frac{1}{2} - \sum_{p=1}^{k_i} \delta_{ip}\right)} = n^{\left(\frac{1}{2} - \sum_{p=2}^{k_i} \delta_{ip}\right)} \tag{8}$$

hence,

$$n^{\left(\frac{1}{2}-\sum_{p=2}^{k_i}\delta_{ip}\right)} < \frac{\sqrt{n}}{4}.$$

Dividing both terms by $n^{\delta_{i1}}$, it follows:

$$n^{\left(\frac{1}{2}-\sum_{p=1}^{k_i}\delta_{ip}\right)} < \frac{n^{\left(\frac{1}{2}-\delta_{i1}\right)}}{4} \tag{9}$$

The first member of relation 9 is, by definition, the size of $Clus(c_{ik_i})$. □

With respect to the cluster forests sequence $\mathcal{F} =< F_1, \ldots, F_{low}, \ldots, F_h >$, we can prove the following:

Lemma 21. *Given the i–th forest F_i such that $i \geq$ low, for at most $n^{\frac{1}{2}-\delta_{i1}}$ successive forests the first cluster could have a size greater or equal than some clusters in F_i.*

Proof. Let us assume, for a contradiction, that

$$|Vert(Clus(c_{j1}))| \geq |Vert(Clus(c_{ik_i}))|$$

where $i \leq j \leq i+t$, with $t \geq n^{\frac{1}{2}-\delta_{i1}} + 1$.

Observe that, cluster $Clus(c_{i+t,1})$ has not be included in all F_j, where $i \leq j < i+t$ even if its size is greater than some other clusters in F_j. This implies that $Clus(c_{i+t,1})$ is not independent with F_j, for all $i \leq j < i+t$. Thus,

$$|Vert(Clus(c_{i+t,1}))| \geq t > n^{\frac{1}{2}-\delta_{i1}}.$$

But this contradicts the hypothesis as $Clus(c_{i+t,1})$ would have been chosen as first cluster in F_i. □

By the above lemma, it derives:

Lemma 22. $|Vert(Clus(c_{t1}))| < \frac{n^{\frac{1}{2}-\delta_{i1}}}{4}$ *where* $t = i + n^{\frac{1}{2}-\delta_{i1}} + 1$.

Proof. The proof follows from lemma 20 and observing that all elements in $Clus(c_{t1})$ are not related to at least one forest F_j, where $i \leq j \leq i + n^{\frac{1}{2}-\delta_{i1}}$. □

From the above technical lemmas it is possible to derive:

Lemma 23. DECOMPOSITION2 *procedure returns an $O(\sqrt{n})$ collection $\mathcal{F} = \{F_i\}$ of forests of clusters.*

Proof. Let $\mathcal{F} =< F_1, \ldots, F_{low}, \ldots, F_h >$ the sequence of cluster forests returned, where F_{low} is the first forest with $k_{low} > 1$. Let us consider the two sub-sequences $< F_1, \ldots, F_{low-1} >$ and $< F_{low}, \ldots, F_h >$. If $k_i = 1$ then $|Vert(F_i)| \geq \frac{\sqrt{n}}{4}$, as by lemma 19 condition 2 holds. Hence, the sub-sequence $< F_1, \ldots, F_{low-1} >$ is $O(\sqrt{n})$.

For $low \leq i \leq h$, by lemma 21 it derives that, for each forest F_i at most $n^{\frac{1}{2}-\delta_{i1}}$ successive forests are composed with clusters having a size greater or equal than at least one cluster in F_i, and, by lemma 22, for F_t, where $t = i + n^{\frac{1}{2}-\delta_{i1}} + 1$, we have:

$$|Vert(Clus(c_{t,1}))| \leq \frac{1}{4}|Vert(Clus(c_{i,1}))|.$$

This implies,

$$h - low \leq \sum_{low}^{h} \frac{n^{\frac{1}{2}-\delta_{i1}}}{4^i} \leq \sqrt{n}$$

□

It is now possible to prove the main theorem of section 1.

Theorem 1. *Let $\mathcal{L}(\prec, \mathcal{N})$ a partial lattice, where $n = |\mathcal{N}|$, a $O(n\sqrt{n})$ -space* implicit *data structure exists, allowing to test partial order relation in $O(1)$-time and to retrieve a total order on the Hasse Diagram in $O(l)$-time, where l is the size of the returned total order. The overall pre-processing time is $O(n^2)$.*

Proof. The data structure used is the one described in section 3, as consequence, lemma 15 gives the desired time bounds for both operations.

In order to prove the claim we have to show that DECOMPOSITION2 procedure returns a decomposition yielding $O(n\sqrt{n})$ -space data structure. Recall that, from lemma 15, the space complexity is $O(nt + nh + \sum_{i=1}^{h} m_i k_i)$, where t is an upper bound on the number of double-trees in a cluster; h in the total number of forests; m_i in the number of vertices connected to forest F_i; and k_i is the number of clusters in F_i. The following assertions hold:

- t is $O(\sqrt{n})$ from base definition;
- $m_i k_i$ is $O(n)$ from termination condition;
- h is $O(\sqrt{n})$ from lemma 23 □

6 Conclusions and Open Problems.

In this paper, a general technique for partial lattices representation has been presented. This technique, based on a two-level graph decomposition strategy, turns to be very efficient, from either space and time complexity point of view, for relation testing problem. Note that, the complexity bound we derived are optimal as they match the theoretical lower bound for this problem.

It is important to underscore that the class of partial orders under investigation has been widely studied [3] and has applications in many fields, such as, for example, the set containement problem and computational geometry.

Moreover, as it turns out from our current research, the decomposition strategy and the data structure introduced in this paper are easy enough to be positively extended for dealing with the dynamic case.

A natural direction for further work is to adopt the same strategy for coping with general partial orders. In fact, the main problem for a straightforward application of the proposed decomposition strategy to general posets is due to the fact that they usually violate lemma 6. In fact, given a poset $P(\prec_P, N)$, and a cluster $Clus(c)$, let u be a vertex in $P(\prec_P, N) - Clus(c)$, if $Clus^+(c) \cap Clus^+(u) \neq \emptyset$ then it can have more than one maximal element. Dually, if $Clus^-(c) \cap Clus^-(u) \neq \emptyset$ then it can have more than one minimal element. This affects time complexity for the partial order relation testing. In fact, as there not exist a one-to-one relation between a given cluster and an external element, it is not possible, given a couple of nodes $< x, y >$, to univocally identify a double-tree containing both x and y.

Nevertheless, the proposed decomposition strategy represents, in this case, an heuristic method for realtion testing which takes into account the sparseness of the Hasse Diagram.

References

1. R. Agrawal. Alpha: an extension of relational algebra to express a class of recoursive queries. In *IEEE 3rd Int. Conf. Data Engineering*, 1987.
2. R. Agrawal, A. Borgida, and H. V. Jagadish. Efficient management of transitive relationship in large data and knowledge bases. In *ACM SIGMOD*, 1989.
3. H. Ait-Kaci, R. Boyer, P. Lincoln, and R. Nasr. Efficient implementation of lattice operations. *ACM Trans. on Prog. Lang. and Syst*, 11:115–146, 1989.
4. H. Alt, K. Mehlhorn, and J.J. Munro. Partial match retrieval in implicit data structures. *Information Processing Letters*, 19:61–66, 1984.
5. Joachim Biskup and Holger Stiefeling. Evaluation of upper bounds and least nodes as database operations. Technical report, ESPRIT-project 311- Advanced Data and Knowledge Management System, 1992.
6. A. Borodin, F.E. Fich, F. Meyer auf der Heide, E. Upfal, and A. Wigderson. A trade-off between search and update time for the implicit dictionary problem. *Theoretical Computer Science*, 58:57–68, 1988.
7. B. Dushnik and E. Miller. Partially ordered sets. *Amer. J. Math.*, 63:600–610, 1941.
8. P. G. Franciosa and M. Talamo. Orders, implicit k-sets representation and fast halfplane searching. *submitted to ORDAL'94*, 1994.
9. G. Gambosi, J. Nesetril, and M.Talamo. On locally presented posets. *Theoretical Computer Science*, 1990.
10. G. Gambosi, M. Protasi, and M.Talamo. An efficient implicit data structure for relation testing and searching in partially ordered sets. *BIT*, 1992.
11. H. V. Jagadish. Incorporating hierarchy in a realation model of data. In *ACM-SIGMOD 1989 Int. Conf. Management of Data*, Portland, Oregon, 1989.
12. T. Kameda. On the vector representation of the reachability in planar directed acyclic graphs. *Information Processing Letters*, 3(3), 1975.
13. D. J. Kleitman and K. J. Winston. The asymptotic number of lattices. *Annuals of Discrete Matemathics*, 6:243–249, 1980.
14. F. P. Preparata and M. I. Shamos. *Computational Geometry.* Springer-Verlag, Berlin, New York, 1985.
15. F. P. Preparata and R. Tamassia. Fully dynamic point location in a monotone subdivision. *SIAM Journal of Computing*, 18(4), 1989.

16. I. Rival. Graphical data structures for ordered sets. In I.Rival, editor, *Algorithms and Orders*. Kluwer Academic Publishers, 1989.
17. R. Tamassia and J. G. Tollis. Reachability in planar digraphs. Technical report, Brown University, Providence, Rhode Island, 1990.
18. Jr W. T. Trotter and Jr. J. I. Moore. The dimension of planar posets. *Journal of Combinatorial Theory*, 22:54–57, 1977.
19. M. Yannakakis. Graph theoretic methods in database theory. In *ACM STOC*, 1990.

Lecture Notes in Computer Science

For information about Vols. 1–751
please contact your bookseller or Springer-Verlag

Vol. 788: D. Sannella (Ed.), Programming Languages and Systems – ESOP '94. Proceedings, 1994. VIII, 516 pages. 1994.

Vol. 789: M. Hagiya, J. C. Mitchell (Eds.), Theoretical Aspects of Computer Software. Proceedings, 1994. XI, 887 pages. 1994.

Vol. 790: J. van Leeuwen (Ed.), Graph-Theoretic Concepts in Computer Science. Proceedings, 1993. IX, 431 pages. 1994.

Vol. 791: R. Guerraoui, O. Nierstrasz, M. Riveill (Eds.), Object-Based Distributed Programming. Proceedings, 1993. VII, 262 pages. 1994.

Vol. 792: N. D. Jones, M. Hagiya, M. Sato (Eds.), Logic, Language and Computation. XII, 269 pages. 1994.

Vol. 793: T. A. Gulliver, N. P. Secord (Eds.), Information Theory and Applications. Proceedings, 1993. XI, 394 pages. 1994.

Vol. 794: G. Haring, G. Kotsis (Eds.), Computer Performance Evaluation. Proceedings, 1994. X, 464 pages. 1994.

Vol. 795: W. A. Hunt, Jr., FM8501: A Verified Microprocessor. XIII, 333 pages. 1994.

Vol. 796: W. Gentzsch, U. Harms (Eds.), High-Performance Computing and Networking. Proceedings, 1994, Vol. I. XXI, 453 pages. 1994.

Vol. 797: W. Gentzsch, U. Harms (Eds.), High-Performance Computing and Networking. Proceedings, 1994, Vol. II. XXII, 519 pages. 1994.

Vol. 798: R. Dyckhoff (Ed.), Extensions of Logic Programming. Proceedings, 1993. VIII, 362 pages. 1994.

Vol. 799: M. P. Singh, Multiagent Systems. XXIII, 168 pages. 1994. (Subseries LNAI).

Vol. 800: J.-O. Eklundh (Ed.), Computer Vision – ECCV '94. Proceedings 1994, Vol. I. XVIII, 603 pages. 1994.

Vol. 801: J.-O. Eklundh (Ed.), Computer Vision – ECCV '94. Proceedings 1994, Vol. II. XV, 485 pages. 1994.

Vol. 802: S. Brookes, M. Main, A. Melton, M. Mislove, D. Schmidt (Eds.), Mathematical Foundations of Programming Semantics. Proceedings, 1993. IX, 647 pages. 1994.

Vol. 803: J. W. de Bakker, W.-P. de Roever, G. Rozenberg (Eds.), A Decade of Concurrency. Proceedings, 1993. VII, 683 pages. 1994.

Vol. 804: D. Hernández, Qualitative Representation of Spatial Knowledge. IX, 202 pages. 1994. (Subseries LNAI).

Vol. 805: M. Cosnard, A. Ferreira, J. Peters (Eds.), Parallel and Distributed Computing. Proceedings, 1994. X, 280 pages. 1994.

Vol. 806: H. Barendregt, T. Nipkow (Eds.), Types for Proofs and Programs. VIII, 383 pages. 1994.

Vol. 807: M. Crochemore, D. Gusfield (Eds.), Combinatorial Pattern Matching. Proceedings, 1994. VIII, 326 pages. 1994.

Vol. 808: M. Masuch, L. Pólos (Eds.), Knowledge Representation and Reasoning Under Uncertainty. VII, 237 pages. 1994. (Subseries LNAI).

Vol. 809: R. Anderson (Ed.), Fast Software Encryption. Proceedings, 1993. IX, 223 pages. 1994.

Vol. 810: G. Lakemeyer, B. Nebel (Eds.), Foundations of Knowledge Representation and Reasoning. VIII, 355 pages. 1994. (Subseries LNAI).

Vol. 811: G. Wijers, S. Brinkkemper, T. Wasserman (Eds.), Advanced Information Systems Engineering. Proceedings, 1994. XI, 420 pages. 1994.

Vol. 812: J. Karhumäki, H. Maurer, G. Rozenberg (Eds.), Results and Trends in Theoretical Computer Science. Proceedings, 1994. X, 445 pages. 1994.

Vol. 813: A. Nerode, Yu. N. Matiyasevich (Eds.), Logical Foundations of Computer Science. Proceedings, 1994. IX, 392 pages. 1994.

Vol. 814: A. Bundy (Ed.), Automated Deduction—CADE-12. Proceedings, 1994. XVI, 848 pages. 1994. (Subseries LNAI).

Vol. 815: R. Valette (Ed.), Application and Theory of Petri Nets 1994. Proceedings. IX, 587 pages. 1994.

Vol. 816: J. Heering, K. Meinke, B. Möller, T. Nipkow (Eds.), Higher-Order Algebra, Logic, and Term Rewriting. Proceedings, 1993. VII, 344 pages. 1994.

Vol. 817: C. Halatsis, D. Maritsas, G. Philokyprou, S. Theodoridis (Eds.), PARLE '94. Parallel Architectures and Languages Europe. Proceedings, 1994. XV, 837 pages. 1994.

Vol. 818: D. L. Dill (Ed.), Computer Aided Verification. Proceedings, 1994. IX, 480 pages. 1994.

Vol. 819: W. Litwin, T. Risch (Eds.), Applications of Databases. Proceedings, 1994. XII, 471 pages. 1994.

Vol. 820: S. Abiteboul, E. Shamir (Eds.), Automata, Languages and Programming. Proceedings, 1994. XIII, 644 pages. 1994.

Vol. 821: M. Tokoro, R. Pareschi (Eds.), Object-Oriented Programming. Proceedings, 1994. XI, 535 pages. 1994.

Vol. 822: F. Pfenning (Ed.), Logic Programming and Automated Reasoning. Proceedings, 1994. X, 345 pages. 1994. (Subseries LNAI).

Vol. 823: R. A. Elmasri, V. Kouramajian, B. Thalheim (Eds.), Entity-Relationship Approach — ER '93. Proceedings, 1993. X, 531 pages. 1994.

Vol. 824: E. M. Schmidt, S. Skyum (Eds.), Algorithm Theory – SWAT '94. Proceedings. IX, 383 pages. 1994.

Vol. 825: J. L. Mundy, A. Zisserman, D. Forsyth (Eds.), Applications of Invariance in Computer Vision. Proceedings, 1993. IX, 510 pages. 1994.

Vol. 826: D. S. Bowers (Ed.), Directions in Databases. Proceedings, 1994. X, 234 pages. 1994.

Vol. 827: D. M. Gabbay, H. J. Ohlbach (Eds.), Temporal Logic. Proceedings, 1994. XI, 546 pages. 1994. (Subseries LNAI).

Vol. 828: L. C. Paulson, Isabelle. XVII, 321 pages. 1994.

Vol. 829: A. Chmora, S. B. Wicker (Eds.), Error Control, Cryptology, and Speech Compression. Proceedings, 1993. VIII, 121 pages. 1994.

Vol. 831: V. Bouchitté, M. Morvan (Eds.), Orders, Algorithms, and Applications. Proceedings, 1994. IX, 204 pages. 1994.